AF871475

A GUIDE TO ASSET MANAGEMENT: WHAT, WHY, AND HOW

Heather Himmelberger, PE

2026

Water Environment Federation
601 Wythe Street
Alexandria, VA 22314–1994 USA
https://www.wef.org

Copyright © 2026 by the Water Environment Federation. All Rights Reserved. Permission to copy must be obtained from WEF.

Water Environment Research, WEF, and WEFTEC are registered trademarks of the Water Environment Federation.

ISBN 978-1-57278-434-5

IMPORTANT NOTICE

This peer-reviewed publication is intended to provide information through a review of technical practices and detailed procedures that research and experience have shown to be functional and practical.

The material presented in this publication has been prepared in accordance with generally recognized engineering principles and practices and is for general information only. This information should not be used without first securing competent advice from an industry professional with respect to its suitability for any general or specific application.

No reference made in this publication to any specific method, product, process, or service constitutes or implies an endorsement, recommendation, or warranty thereof by WEF.

WEF makes no representation or warranty of any kind, whether expressed or implied, concerning the accuracy, product, or process discussed in this publication and assumes no liability.

Anyone using this information assumes all liability arising from such use, including but not limited to infringement of any patent or patents.

WEF works hard to ensure that the information in this publication is accurate and complete. However, it is possible that the information may change after publication, and/or that errors or omissions may occur. We welcome your suggestions on how to improve this publication and correct errors. WEF disclaims all liability for, damages of any kind arising out of use, reference to, or reliance on information included in this publication to the full extent provided by state and federal law.

About WEF

The Water Environment Federation (WEF) is a not-for-profit technical and educational organization of more than 30,000 individual members and 75 affiliated Member Associations representing water quality professionals around the world. Since 1928, WEF and its members have protected public health and the environment. WEF and its global network of members and member associations provide water quality professionals around the world with the latest in water quality education, training, and business opportunities. WEF's diverse membership includes scientists, engineers, regulators, academics, utility managers, plant operators, and other professionals. WEF uses this collective knowledge to further a shared goal of improving water quality around the world. To learn more, visit www.wef.org.

Contents

List of Figures

List of Tables

Glossary

Asset Management: Water utilities have to manage their assets regardless of whether they take on a formal asset management program or not. They need to make decisions about operations, maintenance, repair, rehabilitation, replacement, workforce, and many other things. What is referred to as "asset management" in this book is a formal, strategic, structured approach to managing assets. Systems may employ some portions of strategic asset management currently but may not have built them into a full, robust, structured program. The use of the words in this book such as "asset management," "asset management program," and "robust asset management," are meant to convey the specific, formal, structured type of asset management that is promoted and, in some cases, required by regulatory or funding agencies in the United States. The book is not trying to discount current activities but rather help utilities build them into a more efficient and effective way to manage assets, one in which employees, elected leaders, and customers benefit from the improvements.

Water: When water is used in the sense of water utility or water system, it is meant in the broadest sense of water, meaning drinking water, wastewater, or stormwater. If water is meant in the more limited way of drinking water, it will be specified as such. In most cases water is used to cover all three of these functions to avoid having to write out drinking water, wastewater, and stormwater each time this is meant.

Utility: The terminology "utility" is used to mean any type of water system and is not meant to be a privately owned entity. It is meant as any ownership type—municipal, homeowner, private, authority, district, or any other. Further, water utility and water system are used interchangeably and in this book are meant as the same thing.

System: See Utility. System and utility are used interchangeably.

Organization: If the term "organization" is used, it is meant to be the water system, utility, municipality, or other organizational structure of the water system. It is generally just another term for the water system.

Elected Leader: "Elected Leader" is being used to signify the ultimate decision-maker(s) of a utility/system. This term refers to any type of water leader whether the person is actually elected or is an appointed volunteer, owner, or any other type of ultimate decision-maker.

Community: In some cases in the book, the word "community" is used. This term is meant to convey the water system (or water utility) and its surrounding areas. It substitutes in most cases for the term system or utility to allow for additional words to be used.

List of Acronyms

AC	Asbestos Cement
AM	Asset Management
AM IQ	Asset Management Intelligence Quotient
AMP	Asset Management Plan
ArcGIS	ESRI's name for its Proprietary Geographic Information System Software
ARRA	American Reinvestment and Recovery Act
AWIA	America's Water and Infrastructure Act
AWWA	American Water Works Association
BCE	Business Case Evaluation
BIL	Bipartisan Infrastructure Law
BMP	Best Management Practice
BOD	Biochemical Oxygen Demand
Cap Ex	Capital Expenses
CCR	Consumer Confidence Report
CCTV	Closed Circuit Televising
CDBG	Community Development Block Grant
CDC	Centers for Disease Control
CIP	Capital Improvement Plan
CMMS	Computerized Maintenance Management System
COF	Consequence of Failure
CPI	Consumer Price Index
CSOs	Combined Sewer Overflows
EFC	Environmental Finance Center
EFCN	Environmental Finance Center Network
EPA	Environmental Protection Agency
ESRI	Environmental Systems Research Institute
FEMA	Federal Emergency Management Agency
FOG	Fats, Oils, Grease
FTEs	Full Time Equivalents (for staffing based on a full-time employee, 2080 hours per year)
FY	Fiscal Year
GAO	Government Accountability Office
GI	Green Infrastructure
GIS	Geographic Information System
GPS	Global Positioning System
GSI	Green Stormwater Infrastructure

HDPE	High Density Polyethylene
HVAC	Heating, Ventilation, and Air Conditioning
ID	Identification
I/I	Infiltration and Inflow
IIJA	Infrastructure Investment and Jobs Act
IT	Information Technology
KISS	Keep It Simple and Sustainable
Lat/Long	Latitude/Longitude
LCRR	Lead and Copper Rule Revisions
LOS	Level of Service
M	Million
MCL	Maximum Contaminant Level
MGD	Million Gallons per Day
MH	Manhole
mg/L	Milligrams per Liter
MTBF	Mean Time Between Failures
MTTR	Mean Time to Repair
NPDES	National Pollutant Discharge Elimination System
NPDWS	National Primary Drinking Water Standards
O&M	Operation and Maintenance
Op Ex	Operating Expenses
PER	Preliminary Engineering Report
POF	Probability of Failure
POTW	Publicly Owned Treatment Works
PPE	Personal Protective Equipment
PRV	Pressure Reducing Valve
PSI	Pounds per Square Inch
PVC	Polyvinyl Chloride
QA/QC	Quality Assurance/Quality Control
RCM	Reliability Centered Maintenance
RD	Rural Development
RF	Redundancy Factor
RFP	Request for Proposals
RO	Reverse Osmosis
SCADA	Supervisory Control and Data Acquisition
SMART	Specific, Measurable, Attainable, Relevant, and Time-Bound
SOPs	Standard Operating Procedures
SRF(s)	State Revolving Fund(s)
SW EFC	Southwest Environmental Finance Center
SWP	Source Water Protection
S2S	Storm to Shade
TSS	Total Suspended Solids

TV	Television
US	United States
USDA	United States Department of Agriculture
UV	Ultraviolet
VFD	Variable Frequency Drive
WEF	Water Environment Federation
WIFIA	Water Infrastructure Finance and Innovation Act
WSU	Wichita State University
Y/N	Yes/No
3-D	Three Dimensional
24/7/365	All the time/every single day of the year

Preface

In Trever Noah's book, *Born a Crime*, he talks about the old saying, "if you give a man a fish he has a meal and if you teach a man to fish he can feed himself for a lifetime" and how it leaves a very important component out. His insight is that the man not only needs to learn how to fish, he also needs a fishing pole. As I was writing this book, I listened to the audio version of Trevor Noah's book (which he narrates) and was struck by both the simplicity and the importance of what he was saying. Because my mind is always pondering asset management, no matter where I am or what I am doing, I saw his observation as an important component of asset management.

As someone who routinely encounters water and wastewater professionals in a variety of settings, I'm often asked to solve a particular problem someone is having. That is similar to giving a man a fish. I could solve that problem and walk away. However, I usually take the time to describe how a more robust asset management program could not only solve the individual problem, but also could improve the whole operation and management. This is teaching a man to fish. This book and its associated resources will enable a person to develop and implement an asset management program and are intended to provide the fishing rod.

I sincerely hope those who pick up this book are able to use this fishing rod to learn to fish and improve their operation and management. I hope the tools help to improve an employee's workday, build trust between customers and elected leaders, extend the life of assets, and overall lift up the water profession. (As an aside, I used the terminology "man" because that is how the saying is commonly used. As a woman in engineering, especially one who joined the profession when very few women were part of it, I am sensitive to those who wish for a more inclusive term. I mean no disrespect and the word "man" should be taken to mean any person.)

1.0 PURPOSE/USE OF THE BOOK

This book is meant to be educational in nature. It is not intended to serve as a comprehensive reference or manual of practice but rather a narrative guide that explains what asset management is and why it is so important for utilities to engage in it. The book is meant to be accessible to any audience, especially smaller-sized utilities, who may be less familiar with it. Because of the desire for accessibility and understanding at all levels, it is written in a less formal manner. The second person voice is used in some cases to help draw the reader in.

The book is meant to be understandable by any person associated with a water (drinking water, wastewater, stormwater) utility/system, whether they are in the position of an elected leader, executive management, management, field staff, or even an interested customer. Everyone at a utility should be involved in the asset management program if maximum benefits are to be achieved. Therefore, this book is meant to provide information to all levels of personnel. It also describes the need to share information about asset management with customers.

Asset management is a thought process that can be considered applied common sense. It should not feel foreign, uncomfortable, or overly sophisticated. Asset management can be applied at any level to achieve the goals the utility sets. The book is intended to convey the simplicity that underlies the strategic asset management approach.

The hope is that a user can read the book, be inspired to take on asset management, begin wherever is the most comfortable, and then expand efforts from there. Other resources can be used to supplement the material in the book and help personnel or utilities as a whole to improve their practices.

The book was written in a particular order to flow from one part of asset management to the next but a reader may choose to select various chapters in a different order to meet their own needs. However, the first chapter is important to read first, as it sets the tone for asset management overall and provides context for the materials in the remaining chapters. It is also important to read the "Terminology Used" section to understand how various terms are used in the book.

A few concepts that should be clarified are the idea that people are assets and should be thought of as such in an asset management program and that everyone engaged in asset management should strive for world's best practice. People are absolutely critical to the development and implementation of asset management and it will NOT succeed without properly trained, skilled staff members. Because of the importance of personnel, some practitioners in asset management want to view staff as assets, and in some cases, see staff as the most important asset. In both cases, the importance of staff is recognized; the difference is whether they should be classified as "assets." This book chooses to recognize the vital importance of personnel but not categorize them as assets.

While it may seem that seeking a level of world's best practice ought to be the goal of all entities who employ asset management, that is not necessarily in everyone's best interest. The costs of achieving this level may outweigh the benefits, especially for medium and small systems. Rather, it is important to "right-size" asset management practice to achieve the levels necessary to meet customer desired levels of service; improve asset longevity, system sustainability, and operator satisfaction and safety; and ensure the system is properly funded. A utility is likely to set different goals for maturity level they want for different parts of asset management. One area may be close to world's best while others may be mid-level or lower. A utility can also change their desired level of maturity over time as they get more comfortable with the practice. The intent of this book is to help people get started, find some early successes, and build a program from there. It is not trying to push everyone to the highest levels of practice. There are resources, such as Australasia's International Infrastructure Management Manual (often referred to as IIMM), that are well-suited to advancing towards world's best practice levels.

2.0 GOALS

The goal of asset management is to improve how we take care of the extensive assets that make up drinking water, wastewater, and stormwater facilities across the United States. It should not be seen as one more thing added to the workload. Because workloads are already too burdensome, asset management may come across as something extra that there is no time for. While there may be extra work initially to create an asset management program and potentially some investment in technology, equipment, or additional personnel, over time asset management will improve the workday, make the utility more efficient, and support necessary rate increases and capital investments. Utilities are already managing assets; this approach just improves those management activities. It is not trying to increase activities, but rather have personnel do different activities or do activities in a more efficient way.

Readers are encouraged to take the time to read the chapters thoroughly and internalize the thinking part of asset management to help carry you on your journey. Tools and technology can be added over time; the thinking is most important. Please keep an open mind to the possibility of change, not for the sake of change, but to improve daily operations, management, and long-term efficiency and sustainability.

Enjoy the journey!

Chapter 1

Why Asset Management?

Rather than, "Why Asset Management?" the question should probably be why NOT asset management? Water utilities (drinking water, wastewater, and stormwater) must manage infrastructure every single day to ensure safe, clean water enters the home and returns to the environment. Drinking water treatment and delivery and wastewater collection and treatment are the foundation of all communities. Additionally, stormwater management services keep communities from flooding, return stormwater to the environment, and improve its quality. Ensuring these vital services are provided at the level desired by customers is essential. Asset management is a strategic thought process meant to help utilities better manage their infrastructure. It is not intended to make someone work harder or more, but rather to work better with more efficiency and effectiveness.

In one example, a medium-sized utility was collecting data throughout its wastewater treatment plant and writing the results—the exact same information—on paper in eight different locations. It was recorded in all those areas because multiple people needed it. Imagine the time it took to walk around the plant each day and write the information down. Now, consider an alternative in which information is entered once into a digital format that is accessible to all. Because it took approximately 30 minutes every day, and the new approach took 5 minutes, a total of 152 hours per year was being spent inefficiently. The utility gained 3.8 weeks of time over the course of the year that can be used some other way on a more important task.

In another example, utility personnel were asked to conduct inspections of all the pump stations using a multi-paged paper form. When an asset management program was started, the form was evaluated to determine how much of the information was considered useful by someone else in the facility. Approximately half of the information was deemed obsolete or unnecessary. Some of the other data was in a format that was not valuable or only told part of the story (i.e., knowing that the pump is running is less important than knowing the run time of the pump). A paper logbook was included in each pump station that contained valuable information about alarm conditions, weather, any unusual situation that occurred, and anything else deemed important. This data could be extremely valuable if it was in a digital format of some type that everyone could access and which could be used to trend information such as what type of alarm conditions occurred and when. In fact, when the utility did start its asset management program, they began digitally collecting pump station information, including logbook data. They determined a particular alarm occurred after a given set of circumstances and were able to make changes that resulted in an elimination of that relatively frequent alarm condition. Collecting only useful data during pump station inspections saved time and effort and analyzing alarm data eliminated overtime pay related to that event. These examples are but a few of the numerous opportunities there are across a utility to improve asset management. If several of the inefficiencies can be improved and the time savings added up over the course of a year, the utility will have the opportunity to address many more critical items and better serve the customers. Operators will have a more satisfying workday and see the connections between the activities they do and the results.

Because asset management is the main subject of this book, a quick definition of an *asset* and *management* is an important starting point. An asset is anything someone owns that has real or perceived value. There are many more nuances of this definition covered in Chapter 3, but this definition should suffice for this discussion. The term management is meant to convey all the actions taken to ensure the system functions now and sustainably into the future to meet the needs of the customers. This term will be discussed in greater detail in Chapters 4 and 5.

A few examples of the difference between current practices, without a direct consideration of asset management practices and principles, and one that does include implementing asset management practices, are presented below.

- *Finding Assets for Operation, Maintenance, Repair, Emergencies:* Not all assets are visible or have known locations. Some may only be known to a limited number of personnel, or only one person, at a drinking water or wastewater utility; others may have been buried under pavement, located in an unusual spot, or have been forgotten over time. If these assets need to be located for any reason (operations, maintenance, repair, rehabilitation, replacement, or because work is being done near the asset by a construction crew), it may take considerable time to locate the asset. Every minute spent looking is inefficient time. It does not improve the asset or situation in any way. The investigation can be more difficult if personnel are looking for the asset in bad weather—flooding, rain, snow—or if the situation is getting worse every minute spent looking for the asset (e.g., a leak of water that keeps flooding the area while the pipe or valve is being located). If assets can be located on a map, schematic, digital archive, or some other systematic approach that is easily accessible to all employees, this inefficient time can be reduced or eliminated allowing the personnel to spend their time on the actual operations, maintenance, repair, or replacement. (Current State of the Assets component of asset management, Chapter 3)

- *Reactionary Operation:* When systems move into reactionary operation, most of the day is spent addressing assets that are failing or have already failed. Little time is available to do preventative maintenance or consider how operational practices may improve asset maintenance and vice versa. The failing or failed assets may disrupt the planned activities for the day because that asset cannot wait and the needs of the failed asset take precedence. When this situation happens repeatedly, it can become frustrating, expensive, and inefficient. It may require personnel to walk past an asset that is showing signs of trouble (e.g., noise, vibration, leakage) to get to one that is even worse. The problem asset that the personnel are walking past today, which could be addressed while it is a minor issue, will be likely be the problem that has to be addressed as an emergency in a few days, weeks, or months. Addressing assets in this reactive manner is more expensive and can even be less safe for staff. With asset management, the emphasis is on more proactive activities, less reactionary ones, and creating a better balance between work done in a proactive manner and reactive manner. No system will ever be 100% proactive (and it would actually be too expensive and in some ways less efficient than one that is well-balanced between the two). (Life Cycle Costing component of asset management, Chapter 5)

- *Using the Information Gathered at the System:* Water systems may gather considerable information about their day-to-day activities, such as daily checklists at a pump station or treatment unit, but the information may never be turned into actionable data because it may be on pieces of paper or written in a logbook. Actionable data allows a utility to examine trends, sort or search the data, review the effect of activities on the condition, risk, or useful life of the assets, and assign work orders to underperforming assets. Without changing the information into knowledge, the utility is operating at a great disadvantage. The information is there to identify a problematic situation, but there is little chance of identifying it because the information is too disparate. (Current State of the Assets and Life Cycle Costing components of asset management, Chapters 3 and 5, respectively)

- *Understanding and Providing the Customer's Desired Level of Service:* A utility's goal is to provide service to its customers. Meeting regulatory compliance is one measure of whether the service is at an acceptable level, but it will only address a small portion of what a utility does. The rest of the service levels—such as how customer complaints are addressed, how quickly a pipe is fixed, the operating pressure, the number of sewer backups, the quantity of non-revenue water and infiltration/inflow tolerated—are able to be set between a utility and its customers. The target levels chosen will affect costs, customer satisfaction, sustainability, resilience, reliability, and many other factors. It is, therefore, very important that the utility meet as closely as possible the needs of the customers. Target levels that fall short of customer expectations may lead to complaints, dissatisfaction, and unwillingness to pay rates. Target levels that far exceed customer requirements may cause the utility to over-expend funds in one area at the expense of another. Understanding customer needs/desires and matching them to the extent possible creates customer support for the utility and helps ensure necessary rate increases are supported. (Level of Service component of asset management, Chapter 2)

- *A Utility's Needs Are More Than Just Short-Term:* Most utilities have some type of capital improvement plan (CIP) for asset replacements but often these are only short-term, covering 10 years or less. The problem with a short-term plan is the belief that at the end of the plan, there is no additional need for funding. For example, if a plan is proposed for the next 5 years, it may appear that years 6 through 20 do not have any requirements for money. The temptation may exist to push off projects on the short-term plan assuming that it is okay to spread out 5 years of funding over 7 or 10 years. While intuitively the decision-makers may know that some amount of funding is required on an ongoing basis, without a long-term plan revealing the level of funding required, the temptation to put off necessary capital improvements may be too great. Without adequate information, a decision-maker may not understand that a very large capital project is looming in year 10 because projects have been deferred for a long time and further delaying short-term problems will only make the situation worse. It is advisable for a utility to have both short- and long-term capital improvement plans with the short-term covering 10 years or less and the long-term showing estimated expenses for 50 to 100 years. Clearly, the level of detail in the short-term plan is higher than the long-term plan, with the closest years (years 1 to 3) having the most detailed information, but understanding the need for replacements and potential large spikes in replacement funding is extremely important for decision-makers. (Long-Term Funding component of asset management, Chapter 6)

- *Avoiding Catastrophic Failures:* Not all assets are equally critical to the operation of the utility and there are some that keep an operator or manager up at night thinking, "I hope X or Y or Z does not fail/break tonight." These are the assets that are close to failure and the consequences if they do fail are high. Some events can even be catastrophic in which many other assets (e.g., roads, buildings, sidewalks, other utilities, or personal property) are damaged when the asset fails. Or, alternatively, the failure prevents the utility from being able to serve its customers or a portion of its customers for an extended period of time. While a utility may not be able to prevent every catastrophic failure, they want to minimize the number as much as possible. Catastrophic failures can cost tremendous amounts of money, put the health and safety of staff and the community at risk, and reduce the confidence in the utility. Understanding the risk of assets allows the utility to rank assets against each other and ensure that repair, rehabilitation, and replacement of assets, as well as maintenance activities, are focused on reducing overall risk and minimizing the chance of catastrophic failures. (Criticality component of asset management, Chapter 4)

The descriptions above are but a few examples of the benefits of undertaking asset management. Routinely, operators and managers may find themselves in inefficient situations, with too much unusable information, too little worthwhile data, and budgets that do not address the pressing concerns. This has been the situation for decades and it is unlikely that anything dramatic will happen to improve this reality from the outside. Water professionals need to take matters into their own hands and adopt a systematic approach, grounded in common sense and years of experience. That approach is to develop and implement a robust asset management program.

To provide a bit more context, a few real-world stories are shared below along with the ways in which improved asset management practice could have helped the situation.

A water utility was alerted that there was a water main break on a service line near a house. In this case, the road was the high point with the house sloping downward on the property. The driveway also sloped somewhat towards the house. The utility came out and spent considerable time denying that the break was on the city side of the meter, all the while water was pouring out. Eventually, they determined that it was indeed the city's portion of the service line that had failed. The water flow was getting worse, and water was entering the garage. The city was not able to find the valve to shut off the water flow in that area and spent a few hours on that task. Water continued flowing into the house and eventually down the side of the house and into the fully furnished basement. When the valve was finally located, many thousands of dollars of damage had occurred to the house. The crew spent the better part of a day on this one leak. One would hope that this is not the type of customer service the city wishes to provide its customers and spending a day on one service leak is a bit inefficient.

Consider instead a scenario in which the city knew exactly where the valves were located thanks to an asset inventory and could immediately go to the valve and turn it off. Further, the city may have considerable confidence in the valve's ability to work if a valve exercising and maintenance program had been established. The service line could have been repaired much sooner, the household damage could have been avoided, the inconvenience to the customer and resulting repairs could have been prevented, and the overall customer support for the utility would increase.

In one large utility, the maintenance department was working on a completely reactive basis. Almost all maintenance activity was conducted after an asset had already failed or was imminently expected to fail. Assets that were leaking, making noise, vibrating, or otherwise indicating a potential future failure had to be ignored to address the more pressing problem of assets already out of service. Thus, there was a never-ending cycle of worsening reactivity. In one case in particular, an asset that was leaking and making noise needed an oil change to protect the bearings. It would have been very inexpensive to make this oil change and would have saved a much lengthier and costlier repair later, but that did not happen. Instead, this asset failed completely while other broken assets were repaired or replaced. If instead, the maintenance program was revised such that the ratio of proactive (or planned) maintenance to corrective maintenance was approximately 70% to 80% proactive and 20% to 30% reactive, the operators would be able to perform preventative maintenance, and assets could be addressed before the point of failure. This revision would reduce the frustration of the operators and make the overall maintenance program more efficient and cost-effective. It would also lengthen the life of the assets.

Consider the case of a wastewater utility in which the line maintenance crews go to the same spot in the sewer line to flush out buildups in the line every month. The buildups are due to a defect in the line. The flushing process takes a four-man crew over half the day and requires another two-man crew at the wastewater treatment plant to address the large grease balls that flow to the plant. If the issue is not addressed at the treatment plant, the plant may experience an upset condition, possibly even a permit violation. This flushing has been ongoing for more than 10 years and will continue for the foreseeable future. This particular job is frustrating for operators, it is time-consuming, and no matter how well the line is cleaned each month, it will need to be cleaned again the following month. This situation can feel futile and cause the crews to feel that no progress is being made. Consider instead the idea in asset management of defect elimination; instead of fixing the same problem over and over again, address the root cause of the problem so it does not need to be fixed again. In this case, a portion of the line would need to be replaced to fix the immediate issue. Additionally, the utility needs to develop and implement a fats, oils, and grease (FOG) program. Comparing the cost of complete line replacement for this utility to the costs of the annual cleaning, including six crew members (four on the line, two in the plant) and the equipment needed to conduct this cleaning, showed a payback period of 2.5 years, meaning that the utility has spent far more money on cleaning the pipe over the past 10 years (approximately four times the cost) than the replacement. Without a remedy for the pipe, the costs will continue to incur in a cost-inefficient manner. Asset management would encourage fixing the line over the continued maintenance activities. Furthermore, the time that is now being spent by the six crew members amounts to approximately 360 hours per year. Over the 10-year period, this would amount to 1.7 full-time equivalents (FTEs) employees. Imagine what else could have been accomplished with this "additional" manpower.

If we are to become more sustainable in our practice, improve working conditions for employees, and increase customer service and satisfaction, we cannot do it with the same thinking we have used in the past. We need to think differently. How? Keep an open mind and read on to learn how to adopt an asset management mindset.

1.0 ASSET MANAGEMENT: IT'S THE THINKING THAT MATTERS

The most important concept is that asset management is a THOUGHT process not a technology process. While technology provides important tools to aid in the thinking, technology will never replace the thought process. For example, while technology can determine that 10 assets all share the highest risk category (more on risk later), it cannot take in all the various inputs to determine the correct order to complete these projects. For example, one asset might be a pipe that lies under a street, and the mayor has indicated that the street is currently on a list for repaving over the next few years. The pipe project should likely be put off until that repaving takes place. Similarly, a technology program may not be able to review asset performance data and consider whether specific construction techniques lead to better or worse outcomes for assets over time. Technology is not able to move from a fix-the-same-issue over and over again approach to one in which the root cause is investigated, staff brainstorm solutions, and the problem solved.

Asset management involves people and the use of business processes aided by technology. The best approach is to have people consistently using effective and efficient business processes backed up by technology. Then the thinking is informed by data (e.g., data informed decision-making) to achieve the best of both worlds. There are many companies that have created computerized maintenance management systems (CMMS) or asset management software that will help utilities collect, store, and analyze data. These programs are not, by themselves, asset management; they are tools to aid in asset management implementation. While the tools can be very powerful and extremely helpful, they do not replace the need for individuals to learn the asset management concepts for themselves and apply them.

> One common belief is the idea that one simply need to load a bunch of information into a computer program, push a button and out pops all the answers. There is often disappointment that this is not the case. Unfortunately, there are no shortcuts. It is still necessary to plan, design, construct, operate, maintain, repair, rehabilitate, and replace assets no matter what technology is used.

Using robust asset management practices is a bit of a culture change as well. It is necessary to train your mind to think differently about what you do, how you do it, why you do it, and when you do it. Just because something has always been done a certain way does not mean it always has to be done that way. It is time to consider what could be, not what is.

2.0 THE FIVE CORE CONCEPTS

Advancing robust asset management practice is a strategic business process that enables more efficient and effective management of assets. It involves a thought process that examines what assets an entity has, what that entity wants the assets to do, which assets are most critical to doing that, how to cost-effectively enable the assets to do what they need to over their entire life cycle, and how the operation will be funded over the long term. These elements are referred to as the five core concepts of asset management. Figure 1.1 shows a visual representation of the core concepts and the list below contains the name of each concept.

- Level of Service
- Current State of the Assets
- Criticality
- Life Cycle Costing
- Long-Term Funding

Asset management involves behavior change and a new way of thinking about assets. It is a journey, not a destination, and involves committing to the practices of good asset management as the way business is done now and into the future. It recognizes the past as a learning opportunity but does

FIGURE 1.1 Five Core Concepts of Asset Management (Source: Southwest EFC)

not dwell on mistakes or things that cannot be changed. For example, if assets were installed incorrectly 20 years ago, it is necessary to learn what was incorrect about the installation, the best way to fix it going forward, and to implement that new approach for future installations. It does not help to focus a lot of energy on lamenting the poor practices; that is wasted effort that is best spent elsewhere. Sometimes, it is helpful to consider that, most likely, individuals were doing the best they could with the knowledge, materials, equipment, and practices they had at the time. It is highly likely that at least some of the practices used today will be modified or improved upon by operators in the future as part of a continuous improvement effort within an organization.

Because asset management is a living, growing, learning approach, an organization's effort to adopt improved practices is never done. It is expected that an entity that is embarking on this journey will have numerous modifications, additions, deletions, and refinements to its approach over the years. It is best to "just do it" and learn by doing rather than spending considerable time up front trying to refine the asset management approach and then implementing it. It is relatively easy to make adjustments to the program as time goes on.

Some practitioners choose to break down the components further. For example, some have used a 10-step process for asset management that includes the following (Figure 1.2).

The 10 steps are mapped with the five core concepts in Table 1.1. The only one of the 10 steps that does not fit neatly within the five core components is the step titled, "Build Asset Management Plan." This particular step is considered an additional, related activity with the five core components. It is a step that would take place after the entity strategically assessed what it wanted to accomplish with the five core components and would involve writing an Asset Management Plan to communicate what the entity is trying to do and how it plans to get there (See Chapter 8).

There are other models adopted across organizations that promote asset management, including governmental agencies, asset management trade organizations, water and wastewater trade organizations, engineering

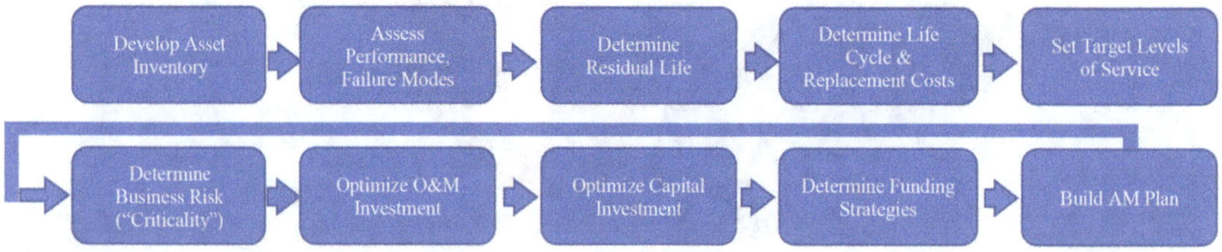

FIGURE 1.2 Ten Steps of Asset Management

TABLE 1.1 Five Core Components and Ten Steps of Asset Management

Core Concept	Ten Steps
Current State of the Assets	Develop Asset Inventory; Determine Residual Life
Level of Service	Set Target Levels of Service
Criticality	Assess Performance, Failure Modes; Determine Business Risk ("Criticality")
Life Cycle Costing	Determine Life Cycle and Replacement Costs; Optimize Operation and Maintenance (O&M) Investment; Optimize Capital Investment
Long-Term Funding	Determine Funding Strategies

professionals, but all would include some version of the five core concepts. Because the Environmental Protection Agency (EPA) is leading the way in terms of federal incentives and mandates through regulation or funding requirements and this agency promotes the five core concepts, this book focuses on this framework.

3.0 QUESTIONS, CONNECTIONS, SOLVING

The best reason to advance asset management practices is for the benefits that will accrue to the entity, its customers, and very importantly, its employees. When employees benefit, their jobs may become more satisfying, their willingness to stay may increase, the system operates more efficiently, and it may be easier to justify funding for specific projects.

One important concept that underlies the ability to help employees with asset management is the idea presented below:

Just because something has been done the same way for years and years does not necessarily mean it should be done that way. It is very important to question beliefs and assumptions. It is equally important to empower all employees to ask questions about activities that seem inefficient or for which there may be a better solution. An employee who does the same job repeatedly has a better vantage point on a particular concern about how the job is being done than someone who is higher up in the management chain.

Asking questions leads to gathering information and data; assessing and analyzing what is currently being done; examining possible options; and selecting the best approach for the future. In some cases, the analysis might show that what is being done is the right thing to do, which can increase confidence in the work activities. If the analysis shows the need for a change and that change is implemented, the improvement in work tasks can reduce frustration, increase efficiency, and create a greater sense of team.

The main point of asking these kinds of questions is to solve issues that employees are having, that are of concern to the system, will improve customer service, or otherwise address inefficiencies and ineffective management. Solving problems will help build support for asset management as employees observe benefits not only for the overall system or its customers but for themselves as well. Solving issues can also reduce the workload for difficult, dangerous, unsafe, or repetitive tasks. The more of these tasks that are eliminated or improved upon, the greater the benefits.

While answering questions and solving issues, it is necessary to seek connections between the components of asset management that can improve the asset management process overall. Seeking connections can include considering the assets involved in solving the issue along with the attribute or other information that would be necessary to conduct the analysis. It might also consider expenditures on operation and maintenance tasks for that type of activity or that set of assets.

Consider one of the previous examples shared regarding a sewer line that was cleaned monthly over a long period of time. The investigation of alternative approaches began when one employee asked the question: "Why do I have to do the same job over and over again? It's an awful job and it is frustrating to do the same thing all the time without making progress. Isn't there something else we could do?"

Starting with this question, what are some connections that can be made to solve the issue?

- *Level of Service:* Is there any level of service goal that relates to pipe cleanings or this pipe in particular, such as a goal related to sewer overflows or the number of times customers are inconvenienced by street blockages? Is there a goal that could be written to relate to this particular issue or issues of this type? For example, "For any sewer requiring cleaning more frequently than every 6 months, a cost-effectiveness study will be undertaken to determine whether there is a better alternative than cleaning. Studies will be repeated once every 5 years on individual pipes if cleaning continues beyond the first study."

- *Assets:* The assets involved in this type of activity include: manholes (necessary to insert the cleaning equipment); pipe segments; trucks; and influent tank at the wastewater treatment plant that receives the waste. Attribute information that is useful to the analysis includes: diameter, material type, slope, the predicted useful life, the number and type of defects in the pipe, length of pipe, depth of bury, replacement cost, installation date, and method of construction when initially installed.
- *Criticality:* The criticality of the pipe in question, both its probability of failure and its consequence of failure, is important information. The pipe has already failed from a level of service standpoint but since it is still able to convey wastewater to some extent, it has not had a complete physical failure. It would potentially rank a 4 on a 1 to 5 probability of failure scale. The consequence of failure would be relatively high as well given that the clogging of the sewer could lead to backups in the sewer system, potentially regulatory violations, and inconvenience to customers. The consequence would rank a 4 out of 5. The overall criticality is 16, which is moderately high. This criticality would indicate a strong need to replace the pipe in the not-too-distant future.
- *Life cycle costing:* Information related to the frequency of cleanings, the cost, the length of time spent, the number of employees involved, any prior repairs to the pipe, the cost of pipe replacement, the potential length of pipe from manhole to manhole, and whether any additional pipe above and below the defect has defects that would warrant replacing a larger section of pipe would all be necessary to make a determination of the best approach.
- *Long-term funding:* Before a full analysis, it is important to determine if the pipe is already included in any Capital Improvement Plan for any reason and if so, in what year? Then it is important to determine if there is money in the budget for annual or other routine pipe replacement and if so, why this pipe has not been prioritized before. Is there enough money in the capital budget to complete this pipe replacement or is there any reserve account funding that could be used to pay for it? If not, will outside funding be sought for this project and any other capital needs that should be completed at the same time?

As this information is gathered and assessed, an analysis can be made regarding the option of continued cleaning versus replacement of the pipe. If the costs are favorable, and the benefits outweigh the cost, the pipe should be repaired/replaced. If the benefits do not outweigh the costs or payback is excessively long (say 30 to 50 years), cleanings should continue.

In this particular case, the payback period was approximately two and a half years, which indicates that benefits of pipe replacement far outweigh the costs. Benefits include: elimination of the monthly cleaning, reduction in potential overflows and regulatory violations, no disposal of removed fats/oils/grease materials (these materials had to be removed at the treatment plant to eliminate potential violations at the plant), and elimination of monthly disruption of traffic. There are also benefits related to opportunity costs by allowing the employees to do more useful tasks and using the trucks for other purposes besides monthly cleaning. (It should be noted that in addition to addressing the defect in the pipe that caused clogs to form and, potentially, backups, another bigger picture part of the solution was to ensure that a pretreatment program that includes addressing fats, oils, and grease at the source [e.g., restaurants, businesses, industries] was developed and implemented. This type of program can reduce the amount of material entering the sewer overall to help reduce problems in the pipe of concern and elsewhere in the system.)

Because this correct solution seems obvious, the costs and benefits are fairly easy to calculate, and the payback is short, one might ask, what keeps people from making these kinds of choices? One big reason is simply never asking the question. If we go back to the idea that many tasks are done simply because it is the way things have always been done and tasks are often handed down from one operator to the next, the lack of questions keeps things status quo.

Another major issue occurs when management or other operators within the organization do not listen to the individuals asking the questions. Sometimes there is a feeling that certain positions are not eligible to ask probing questions or these individuals may self-censor, figuring no one will listen. In fact, all employees should be empowered to ask questions.

When data is not gathered to follow up once the question is asked, the issue may not be resolved. To properly analyze the issue and potential solutions, data needs to be obtained, gathered, and assessed. If this process does not occur, there is the risk the question stays in question form and never gets resolved. This lack of action would have negative consequences on personnel who could see inaction as a reason to stop asking questions and may perceive asset management as a "failure." While that is not actually the case, the perception could have a detrimental effect on the overall implementation of the asset management program.

Failing to view operating expenses (*op ex*) the same as capital expenses (*cap ex*) can also be an obstacle. For asset management to function properly, it is necessary to see op ex and cap ex as equal dollars and to choose the set of actions that best serve the utility no matter whether they fall into cap ex, op ex, or some combination of the two. Systems may be reluctant to move something from one side of the budget to the other. Also, because of the funding approaches used in the United States, where outside funding from state and federal governments—such as the State Revolving Funds (SRF) or United States Department of Agriculture (USDA) rural development funds—can be obtained for capital expenses but not operation expenses, some systems may choose options that favor capital projects over anything that might be operational, to the potential detriment of the utility.

Some examples of generic questions that could be asked within a utility are presented below. These suggestions represent a tiny fraction of the types of questions that could be asked to provide some ideas and prompt deeper thinking.

- Where are all the fire hydrants located? This type of question can be helpful because hydrants are used during emergencies for fire suppression and can be used to improve water quality by flushing. Sometimes fire hydrants get buried behind shrubberies or other vegetation making their location more difficult. See the photos below as one example. In the photo on the left, the hydrant is very difficult to see. The hydrant is shown behind the shrubs in the photo on the right.

- Alternatively, the question can be where is any other asset class located (e.g., valves, pipe, wells, pressure reducing valves, manholes, pump stations, lift stations, sample locations)?
- What is the condition of all the fire hydrants? Similar to location, the condition of hydrants is important because if they are in a non-operational condition, they may not function during emergencies, making it difficult to suppress a fire, endangering lives and structures. If they are not able to function as a means to flush pipes, public health might be affected.

- Similarly, what is the condition of any other asset class or individual asset?
- What is the likelihood that the isolation valves in the system will work when used to address a pipe break or other system problem? This question may include information related to the location and condition of isolation valves, which direction they turn, how many turns it takes to close, and any other pertinent information.
- Are all the meters in the billing system able to be located in the field? Are there some meters that are in the billing system but cannot be found? Are there meters in the field that are not located in the meter inventory? Are there meters that are inoperable or that have been tampered with? Since meters are the "cash registers" for the system, if a meter is missing from the billing system, the utility is not receiving revenue for the provided commodity and service. There are many examples of systems undertaking a meter inventory process and finding discrepancies, including missing meters, incorrect readings, incorrect signals, tampered meters, and other problems.

One utility learned that a large number of meters were missing from the billing inventory in one section of town (approximately 30%). Another utility found that its largest restaurant had a meter that was not in the inventory. Still another utility found that in several cases, two households were connected to the same meter.

- Are operators receiving the training they need to properly operate the system? What type of training is best? Is training included in the annual budget? Is the budget adequate to allow all operators to receive the training they need to: become certified, increase certification levels, understand regulations, practice good safety, understand technology and how to use it, and any other important topics? Operator knowledge and skill is critical to the proper operation and management of a utility.
- Is the utility appropriately staffed? Are there vacancies? In what part of the facility? How much does this affect operation and management? What is being done to address vacancies? Could additional actions be taken?
- How can safety be improved at the facility to reduce injury time and protect employees?
- What is the location of all distributed green infrastructure? Are these assets distributed equitably throughout the community? (Green infrastructure is natural or man-made assets meant to mimic natural

processes and intended to serve a water-related function, such as protection of a water source, treatment of stormwater, reduction of stormwater, reduction of infiltration and inflow. Green infrastructure is quite varied and can include: tree trenches, bioswales, rain gardens, green roofs, and permeable pavement. Natural assets include forested land, riparian buffers, streams, rivers, and lakes, and wetlands.) (See Chapter 7)

Take a moment and think about your own system. What are three questions you could ask about your system? Be creative and think about things in which answering the question would help you in some way. Consider the idea that just because we have always done it this way does not mean it has to be done this way. Question your assumptions. During development of your asset management program, build in answering these and other questions. If you can show wins along the way, it will help build momentum for other asset management activities.

- What are the most common maintenance activities performed at the utility? How much time is spent on these activities? Is there any modification that could be implemented to improve efficiency? Is the timing of these activities based on manufacturer's recommendations, asset data, or is it simply the way that it has always been performed?
- How often are pumps currently lubricated? Is this the optimal frequency of lubrication or would a different frequency be better? Is the correct lubricant being used? Would a different lubricant allow for a longer time between replacements?
- What factors most affect infiltration and inflow (I/I) into the system? What areas should be improved to reduce I/I?
- If there is only enough money to replace a small amount of pipe, which pipe is the highest priority? Does the current process of replacing pipe follow a risk-based approach? Does it consider condition rather than just age?

The questions asked can focus on the things that are the most frustrating to employees, items that feel inefficient or ineffective, actions to make the job easier or safer, improvements to customer service, ways to reduce customer complaints, or just about anything else.

Asking questions is an important and valuable part of asset management, but it does not always happen. There are several potential reasons for this lack, including: a lack of understanding the importance of asking questions, being too tied to procedures of the past, not listening to the staff when they ask questions, upper management not being supportive of the idea of questions, and not being willing to adopt a culture that is accepting of questions.

While questions are very important, it is equally important that answers not be assumed. Data must be gathered to analyze the situation and determine the best outcome. It is important to verify the cost/benefit relationship before proceeding. The necessary data may come from an asset management system if one is in use, other records (paper or electronic), operators, managers, other staff, finance department, contractors, or if it does not already exist, it may need to be developed over time as part of the analysis.

The data and analysis should be shared with staff, especially the staff who asked the question. When results are shared, the staff are more likely to accept the outcome, even if the outcome is to keep the status quo. There may be cases in which the current circumstances do not allow the change to be made right now, but the change can be programmed in for a future time. For example, maybe a large project is being planned for 5 years from now and it is best to wait and include this project with the larger project. Or the analysis may show that a revision to the asset is warranted but an upcoming regulatory change renders the asset obsolete. In this case, it is less efficient to make the change now than to wait until the asset is replaced or removed later on.

When performing assessments of the best alternative, it is necessary to consider the difference between cap ex and op ex. The best asset management practice views these two expenses as equivalent. In the United States, there is favorable funding for cap ex, which can include 0% interest loans, below market rate loans, principal forgiveness, negative interest, or a grant. Op ex almost always comes from customers in the form of user fees, taxes, or similar payments. There may be a temptation under these types of circumstances to choose an option that is more heavily weighted towards cap ex than op ex to take advantage of favorable funding. Consideration of funding may outweigh the correct option from a purely asset management perspective; a utility may choose to employ a replacement approach rather than a repair/rehabilitation/maintenance approach because they are eligible for favorable funding, such as a grant, that would not require users to pay back the funding. Funding agencies do not yet require utilities to prove that replacement is the most efficient option (over continued maintenance) in order to receive funding. This reality must be recognized and considered. However, utilities are strongly encouraged to use the best solution regardless of which budget the money will come from (cap ex or op ex).

Readers are encouraged to think creatively and expansively about questions they could ask about their utility. As an initial exercise, convene a group of individuals who have responsibility for the utility—which could be several people in a large utility at various levels or just a few people in a small utility—and ask them to write down three questions they could ask about the facility. The questions should be ones that would benefit employees, the utility, and customers in some ways. Generally, benefits to one (employees, utility, or customers) will benefit all because the three are connected. Examine the questions and choose a few to work on. Showing positive results will help build support for additional asset management activities.

4.0 ASSET MANAGEMENT TEAM

Development and implementation of a robust asset management program is best completed by an asset management team. It is important for this team to include as diverse a group of people as possible in terms of roles (job within the organization) and management level (i.e., from field staff to middle managers, to upper managers, to executive directors). To achieve the maximum benefits, acceptance and participation at all levels of the organization are required. The field staff are the eyes and ears of the organization and will need to collect asset and performance data on an ongoing basis and share their questions and concerns. Managers can use the data for detailed analysis, including trends in performance, necessary asset replacements or maintenance schedule changes, ability to meet level of service goals, and regulatory reporting. Top-level managers communicate the results to the board/governing body/elected leadership to gain support for capital funding, staff increases, equipment needs, or other requirements. Once it is recognized that asset management cannot be fully accomplished by one individual sitting in the corner, the concept of a team and how important it is for everyone to play a role on that team comes into focus.

Ideally, the team is initiated by the senior management or elected leaders to give the team legitimacy and to show that the leadership is behind the endeavor and expects cooperation and participation from staff. This sponsorship is particularly important because attendees will have to forgo their regular work duties to attend meetings and will need to feel that their participation is worthwhile, encouraged, and expected. Additionally, supervisors will need to support the participation of their staff member(s) and will be more likely to do so if they know asset management is a priority of the people at the highest levels in the organization. The personnel at the highest levels of the organization need to provide the overarching vision for the asset management program. They also need to provide the support and the necessary level of resources (people and money) when the asset management efforts hit snags or major hurdles and to facilitate the cultural, technical, and business process changes that are going to happen as asset management activities are implemented.

Initially, at the beginning stages of asset management, the development team may need to meet often, but over time, the meeting schedule can relax a bit. Monthly is a good starting point for meeting frequency. More often than monthly may be burdensome to staff and less frequently will not allow sufficient time to keep the momentum going and ensure that adjustments can be made if activities do not go as planned. Once asset management practice is imbedded in the organization and the program is running smoothly, quarterly meetings may be sufficient. Meetings should not be less often than quarterly because asset management is a living, breathing practice and if the team rarely meets, it is likely that asset management will stagnate or move to the background and benefits may not be realized or the program may not be sustained.

It will be necessary to have someone as the designated leader of the team. That individual must have the authority to run the team meetings, make assignments, and hold individuals accountable for their activities that will require the endorsement of upper management. The individual does not need to have a position of high authority outside the asset management team; for example, a field staff member can be appointed as the team leader as long as the authority is given to that person by upper management and it is clear that they have the full support.

It is important for upper management personnel to come to at least some meetings to show that they believe participation is important and worthwhile. If these individuals are taking time out of their busy days to come to the meeting, it sends a signal to the rest of the staff that this time expenditure is important. It also shows the backing of the overall asset management initiative. While at the meetings, the upper management should be present and participatory, not focused on other work or their computer or phone. Again, this behavior will send a strong message to the rest of the team and the organization as a whole.

The selection of the team leader should not be taken lightly. This person must be a strong supporter or champion for asset management. There is likely to be some dissent in the employee ranks and if the person cannot stand up to this attitude, the meetings and overall asset management effort will not be successful. The leader also needs to understand asset management, the terminology used, its intended purpose, and be able to explain it to the attendees and other staff. The individual does not have to know everything, but if they cannot even explain the basics well, they are likely to lose the confidence of the other participants and may not be able to drive the organization forward. As a champion for asset management, the leader should also be looking for other champions within the organization (either on the asset management team or not) to work alongside them and help their peers see the value and benefit of asset management. These champions can also help resolve issues, overcome roadblocks, and instill a mentality of "here's how we can do this." If there is a designated asset manager position within the utility, this person may be, but does not have to be, the one to lead the meetings.

> If it is hard to identify a champion for asset management right at the beginning, look for those individuals who quietly embrace the concepts and implement them in their routine activities. It may be possible to build these individuals into the champions the organization needs to implement a robust asset management program and have a strong team leader.

There may be skeptics on the team and it will take a strong person to prevent these individuals from derailing the entire process. Good leadership can use the skeptics to have important dialogue about what the concerns are, why they are leery of asset management or their role in it, and how they can be brought on board. Someone who was initially unconvinced but then saw value in asset management can end up being a powerful ally. The individuals who can really create dissention are those who do not believe in advancing practices or have other personal reasons for resisting improved practice and cannot be convinced of its benefits. They will try to bring people to their way of thinking or sabotage the efforts. Again, leadership skills will be necessary to ensure success and the support of upper management can go a long way in helping with these situations. Fortunately, as advanced asset management practices have been adopted by more organizations, there are fewer people who are trying to derail progress in the sector.

> It is important to be on the lookout for individuals who speak positively about asset management in a public setting or in front of upper management but actively work against asset management in their day-to-day activities. If employees are judged by how well they implement their particular part of asset management, it may be easier to identify these individuals and address the issue.

The team should look for examples of success and publicize these widely, both within the team and outside the team. These examples may include small examples—a particular maintenance change that resulted in either a cost savings or a cost avoidance—or they may be much larger examples—creation of a long-term capital improvement plan covering 100 years, development of level of service goals, making a change based on missing a level of service goal. Either way it is necessary to share the successes widely. This recognition should include individual employees or departments to show that the organization appreciates and recognizes their efforts.

While the team should meet as a large group on a regular, periodic basis, it is also possible to create subcommittees out of the group to accomplish particular tasks. These subcommittees can also include others who are not officially on the asset management team to supplement the knowledge base or skill set needed. When subcommittees are formed, they should be intended to serve a particular purpose for a set amount of time; they should not be created to serve indefinitely. At the initiation of a subcommittee, the overall team should specify the subcommittee's purpose and the expected timeline to completion. Then the subcommittee can be formed and those individuals can determine the best way to accomplish the task. Following the completion of the specified task, the subcommittee should be dissolved.

If it is possible to include active participation in the asset management team as one means of evaluation during annual (or more frequent) employee reviews, this might be one way to ensure that participation continues and is recognized. Not all utilities would have this possibility as it would need to fit within the confines of the specific human resource policies of the organization. A message from the top leader of the organization that they are watching the work of the employees in asset management and have an expectation of active engagement in asset management is another way to aid ongoing participation. This message is a very powerful way to reach an individual, such as a supervisor, who may be reluctant to allow a direct report to spend their time on asset management. Messages such as this help address situations in which middle managers feel threatened by the changes and try to impede an employee's participation. To help allay concerns regarding missing work time from an employee's regular duties, top management can let the supervisor know that they will give the supervisor the latitude, the time, and resources to either delay activities or to complete them with other personnel or other methods so the employee involved in asset management can spend time advancing asset management practices.

Not all utilities will have one or more positions that include "asset management" in their title or for which the job description is mostly or completely related to asset management. A lack of positions with "asset management" in the title is particularly likely for smaller utilities that will need to incorporate asset management roles into existing positions. Furthermore, a utility may decide to move asset management forward without titling the project team an "asset management team." They may feel that building asset management planning and implementation into existing roles and having these roles work together to move the organization forward is a better approach. In this case, perhaps an operations supervisor, an operator, an engineer, and accountant, and an IT specialist work together in their current roles to advance practice in each of their respective areas.

For very small utilities, it will be more difficult to have a team atmosphere. In these cases, the team can include the entire staff and possibly a representative from the elected leadership. In cases in which there is only one employee, time should be set aside each month to consider the asset management program with elected leaders or, if that is not possible, maybe outside stakeholders, such as interested individuals (e.g., a retired engineer or one willing to offer pro bono advice) or customers or others in community leadership outside of the water sector can provide some opportunities for a team of individuals who can discuss ideas and offer assistance with implementation of some of them.

The team can be as big or small as necessary to meet the needs of individual utilities. The most important component is to have the opportunity to share information, discuss ideas, determine the best way to implement activities, and provide support to one another. The group should allow every attendee to be at the same level of importance (i.e., no one's opinion should be counted more than anyone else's) no matter their role within the organization. Other than the leader, the remaining individuals should be considered at the same level when the meeting starts. The team members should also view their role as ambassadors for the asset management program and activities both within and outside the organization.

5.0 REFERENCE

Southwest Environmental Finance Center. (n.d.). *AM/CMMS evaluation and recommendations report*. https://swefc.unm.edu/home/resource/am-cmms-evaluation-and-recommendations-report/

Chapter 2

Level of Service

While asset management discussions most often start with the Current State of the Assets, a better starting place is Level of Service. The main function of a drinking water utility is to provide a desired product (safe, clean, healthy drinking water) to its customers. The main function of a wastewater utility is to clean used water (sewage) to a high quality so that it can be safely returned to the environment. More recently, rather than thinking of wastewater effluent as a waste product, wastewater utilities are being thought of as "resource recovery facilities," meaning that the effluent is not a waste product to be disposed of, but rather a resource that can be reused as irrigation water, process water, or drinking water (either directly or indirectly). Additionally, the function of storm water utilities is to prevent flooding, safely move stormwater to a receiving water body, reduce contaminants, infiltrate the stormwater locally, or some combination of these activities. In all three cases, the water utilities are providing a vital customer service. This service is provided by owning, operating, and/or maintaining assets (in some cases, especially with storm water assets, a utility may operate or maintain assets it does not own). The assets, or certain types of assets, could be eliminated if there was another way to achieve the same level of service. So, the assets can come and go or change over time, but the service remains the core business function. Starting with the Level of Service helps reinforce the concept that it is all about the service and the assets are there to support the ability to provide the service.

In Level of Service, the governing body sets the overall mission for the utility. However, it is important for the governing body to seek input from utility staff at all levels and the customers it serves to ascertain their concerns and expectations when considering the utility's mission. The mission statement shares, in written form, the strategic direction for managerial, operational, and financial decisions. It serves as a foundation and the goals for the utility are derived from the mission statement. These goals define in clear terms what the utility wants to provide for its customers. Each goal will include target levels that can be measured against to see how well the utility is doing towards meeting its goals. The goals can be internal, indicating they are most meaningful to individuals working for the utility, or they can be external, indicating they are meaningful to the general public or elected leaders. The progress towards meeting the goals should be clearly communicated to managers, decision-makers, governing bodies, and the general public. A utility may also wish to develop a vision statement that is aspirational and indicates what the utility would be when it achieved its goals. The vision statements are most often seen for large utilities and represent a bit more complexity so no further discussion will be presented here on vision statements. However, there are many good vision statements that can be found with a quick internet search if a utility wishes to develop one.

The concept of level of service is an underappreciated portion of asset management, quite possibly the most underappreciated. If embraced, level of service can provide significant benefits to the utility as well as its customers. This underappreciation is another reason that level of service is discussed first. It is an attempt to bring this component to the forefront and highlight its relevance.

1.0 THE ROAD MAP

Level of Service represents the road map of how the facility will be operated. Similarly to the way a road map shows multiple options to get from point A to point B, a utility has multiple options regarding how to operate its facility. The most direct route between two points would be the most cost-effective and efficient way to get where you were trying to go. There are many other options available that might take you off the main path and take more time, resources, energy, or divert attention from more important details. The Level of Service is trying to provide this thinking for the utility.

It is intended to focus the utility on the most important functions and evaluate how well the utility is doing towards achieving its intended functions.

As an example, consider a goal around how much pipe is televised in a wastewater collection system. The goal could be stated similarly to the following:

> The utility will use CCTV (Closed Circuit Televising) equipment to proactively inspect 10 miles of pipe per year on a prioritized basis (i.e., this goal applies only to pipe that was not otherwise scheduled for maintenance or repair and should be done in a priority order from most important to least important to inspect).

To determine how this goal provides a road map for the utility, consider the actions the utility needs to take in order to achieve a goal of this type:

- Estimate the time it would take to conduct CCTV work on a per-foot or per-mile basis ($/foot or $/mile).
- Determine if sufficient crew members exist to conduct this work.
- Determine if the CCTV equipment exists and if it is in good working order.
- Develop a prioritized list of pipe based on a consideration of factors that would most likely result in deteriorated pipe (e.g., size, type of pipe, location, slope, soil conditions, proximity to surface or groundwater).
- Create a schedule for when proactive CCTV work can be done in the context of all other work that needs to be done.
- Determine if there are times of year when CCTV work cannot be performed (e.g., in the winter, in the heat of the summer, during wet weather).
- Evaluation technology needs to be able to receive, interpret, and react to the data from the CCTV work.

As the utility thinks through these needs, it will necessarily have to come up with a good solution for how the work can be effectively and efficiently completed.

Next, the utility needs to consider what it means if the goal is not met or if the goal is surpassed. In the case of not meeting the goal, the destination (in the analogy of the road map) was not achieved; the journey stopped short. If the utility is close to the goal, perhaps this is of little consequence, but if the goal is missed by a lot, there could be more serious ramifications. If the goal was missed because of an unusual circumstance that is not expected to be repeated, it may not be necessary to change anything at this time. For example, if the goal was not met because the CCTV truck was undergoing repairs for several months, the following year, meeting the goal may be possible.

In the case of overachieving the goal (i.e., completing 15 miles of pipe instead of 10), there are cost implications. This situation would be similar to taking a detour and traveling further than you wanted to on your journey. It costs time and money, but you still got where you wanted to go. It is important to examine whether there was a downside to conducting more work than the goal specified. The questions to assess the situation include the following:

- Was there some specific circumstance that let the utility do more work this year, such as the presence of temporary additional staff or an outside contractor?
- Was there a lower-than-expected demand for corrective work enabling more proactive work to be completed?
- Were there tasks that didn't get completed while the crews were doing the extra CCTV work? Were these tasks higher or lower in priority than the extra televising?
- Was there sufficient budget in the proactive CCTV line item to complete all the work or did another line item have to be reduced because of the extra work?
- Was there a specific need to have additional pipe assessed?
- Did the benefit of the extra CCTV work outweigh the cost to other sectors of the utility?
- If there is a legitimate need to complete the proactive CCTV work faster, would it be better to change the goal to 15 miles of pipe rather than 10?
- How important is this goal to the overall mission of the utility?

Some goals will have more cost implications than others. For some goals doing more than required might have little effect, while for other goals there might be a significant effect. Similarly, there are goals for which meeting the goal has less effect on the overall quality of the operation than others. The most important thing is to cost-effectively and efficiently provide the services that the customers want and to understand the effect of individual goals.

As the utility creates its own road map, it is attempting to navigate the path that will lead the utility where it wants to go in the most efficient, effective way possible. It will need to consider its goals and associated target levels periodically and make adjustments as necessary.

2.0 MISSION STATEMENTS

The mission statement provides an overarching purpose for the utility and describes what it is trying to accomplish. It shows what is most important to the governing body and communicates it in a clear, concise manner. The mission statement puts all employees on the same page regarding how their actions and day-to-day tasks relate to the overall intent of the utility. It should give the employees a sense of purpose and answer to the question of "why are we doing this?" The mission statement can also communicate to customers where the ratepayer money is being spent.

Mission statements can help prioritize activities and spending. In times of reduced revenues or limited staff, the system needs to focus on those activities that most closely align to the mission statement.

If the water system is part of a bigger organization, such as a county or municipality, its mission statement should be aligned with the overall mission statement for the larger organization. If there are multiple water departments, such as a water, wastewater, and stormwater department, it may be easier to have separate mission statements rather than one larger mission statement covering all three sectors. Additionally, if the water service is divided between plant and field, where treatment is in one department and distribution and collection are in another department, there may be a desire to have mission statements that cover these distinct functions. Whether the organization chooses to have one unified statement or a few statements for specific operating units should be based on which method will provide the most guidance to employees and customers in conveying the mission.

3.0 CHARACTERISTICS OF MISSION STATEMENTS

There are good and bad characteristics of mission statements. The Non-Profit Hub provides examples of good and bad mission statements. Their information can be found at the link below the table. Their list of characteristics are also provided in Table 2.1. While not all of these characteristics apply precisely to water-related mission statements, they are good elements to consider in crafting the statement.

Consider a few examples of water-related mission statements. These mission statements were taken from publicly accessible sites for the water systems at the time the book was prepared. (Links for all mission statements in this chapter are provided in the Reference section.)

TABLE 2.1 Characteristics of Good and Bad Mission Statements

Good Mission Statements . . .	Bad Mission Statements . . .
Use language your customers use	Use jargon, do not understand the audience
Are emotionally stirring	Are logical and cold
Communicate "why"	Communicate only the "what" and "how"
Are concise	Are really long
Contain a single, powerful sentence	Contain a rambling paragraph
Sound good when spoken out loud	Are full of clauses and hard to say
Are memorable	Are forgettable
Surprise	Are dull
Are actionable	Cannot be quantified
Are specific	Are vague

Source. Non-profit Hub https://nonprofithub.org/starting-a-nonprofit/nonprofit-mission-statements-good-and-bad-examples/

> WaterOne: Johnson County, Kansas
> "We provide a safe, reliable, high-quality water supply with superior service and value."

How does this particular mission statement fit with the characteristics of a good mission statement (Table 2.2)?

Overall, this mission statement stacks up reasonably well against the characteristics of good mission statements and can serve the utility well.

Consider Mojave Water Agency's mission statement (Mojave Water Agency, n.d.).

> Mojave Water Agency: San Bernardino County, California
> "Collaboratively manage groundwater basins sustainably, import water responsibly, and address risks proactively using sound science."

How does this particular mission statement fit with the characteristics of a good mission statement (Table 2.3)?

Again, this mission statement stacks up well against the characteristics of good mission statements and can serve the utility well.

Below are a few other examples of mission statements (also taken from publicly available information for the water systems and were current at

TABLE 2.2 Characteristics of Good Mission Statements and WaterOne Mission Statement

Good Mission Statements . . .	WaterOne's Mission Statement
Use language your customers use	Words used would be used by customers. Key words in the statement for customers are: safe, reliable, high-quality, superior service, and value.
Are emotionally stirring	Has some emotional elements, such as superior service.
Communicate "why"	The why is more implied than stated.
Are concise	Statement is concise.
Contain a single, powerful sentence	Is a single, powerful sentence.
Sound good when spoken out loud	Works well when spoken out loud.
Are memorable	Could be memorized by utility employees and remembered by them.
Surprise	Not particularly surprising, but water mission statements have less ability overall to surprise than other types of mission statements. Water is a basic function and the service is pretty similar everywhere.
Are actionable	This mission statement is actionable; actions would be required to meet these elements.
Are specific	Relatively specific but some of the words are somewhat open to interpretation. For example, what one person thinks of superior may be different than what another person thinks of as superior. Similarly, value may be interpreted differently.

the time of publication. Links to the internet websites are included in the reference section).

Pierce County Department of Public Works and Utilities, Sewer Utility: Tacoma, Washington
"To provide customers with high quality, cost-effective sanitary services in accordance with all applicable local, state, and federal permit requirements."

TABLE 2.3 Characteristics of Good Mission Statements and Mojave Water System's Mission Statement

Good Mission Statements . . .	Mojave Water System's Mission Statement
Use language your customers use	Words used would be used by customers. Key words in the statement for customers are: sustainably, responsibly, proactively, and sound science
Are emotionally stirring	Has some emotional elements
Communicate "why"	The why is more implied than stated
Are concise	Statement is concise
Contain a single, powerful sentence	Is a single, powerful sentence
Sound good when spoken out loud	Works well when spoken out loud
Are memorable	Could be memorized by utility employees and remembered by them.
Surprise	Not particularly surprising, but water mission statements have less ability overall to surprise than other types of mission statements. Water is a basic function and the service is pretty similar everywhere.
Are actionable	This mission statement is actionable; actions would be required to collaboratively manage, import water responsibly, and address risks proactively
Are specific	Relatively specific but some of the words are somewhat open to interpretation. For example, what one person thinks of sustainably may be different than what another person thinks of. Similarly, sound science can be interpreted differently by different people.

Leucadia Wastewater District: Carlsbad, CA
"To serve the public by collecting, transporting, recycling, and treating wastewater in a safe, reliable, efficient, cost-effective, and environmentally responsible manner while providing excellent service for our customers."

Northeast Ohio Regional Sewer District: Cleveland, Ohio
"Provide progressive sewage and stormwater management through innovation, fiscal responsibility, and community partnerships."

Central Arkansas Water: Little Rock, AR
"Deliver high-quality, affordable, abundant, dependable water services."

Atlantic City Municipal Utilities Authority, Atlantic City, NJ
"Our mission is to deliver the highest quality potable drinking water in a cost-effective and environmentally responsible manner while working together as one team for our customers."

Verona Stormwater Utility, Verona, WI
"To maintain stormwater facilities to an acceptable and legally required service level providing comparable service to all citizens. To expand public facilities as necessary to accommodate the growth of the City minimizing the cost to existing citizens."

While all eight of the mission statements (the six above and the two previous ones) may share some elements in common, they each represent a different prioritization of activities. In all cases, however, these mission statements are well-served by adopting a culture of continuous improvement in asset management. For example, WaterOne talks about "superior service" (WaterOne, n.d.). The level of service portion of asset management is a key element in demonstrating that superior service is being delivered. Without any goals that contain specific target levels to show what superior service means, the utility would not be able to demonstrate what is being delivered to the public or elected leaders. With regularly measured written goals, including associated target levels that are measured and reported on, they are able to show, with results, that a particular level of service is being provided on a regular basis. The goals can also help show when additional work may be required to improve service if it dips below acceptable levels.

The Pierce County and Leucadia Wastewater District, as well as the Atlantic City Municipal Utilities Authority mission statements mention the words "cost-effective" (Atlantic City Municipal Utilities Authority, n.d.). Asset management is intended to be cost-effective by its definition: "at the lowest life cycle cost." These two mission statements would tie directly to criticality and life cycle costing. Similarly, the words fiscal responsibility in Northeast Ohio's mission statement and "superior value" in WaterOne's mission statement fit this same idea. The Verona Stormwater Utility mentions the need to expand necessary public facilities while minimizing costs (Verona, n.d.).

Mojave Water District's mission statement mentions "managing risks proactively," which ties to the idea of criticality in asset management. Similarly, Central Arkansas Water's mission statement includes "dependable," which ties it to criticality and the condition assessment portion of the current state of the assets (Central Arkansas Water, n.d.).

A note to the readers: the above discussion uses real mission statements from utilities that were found online on their public websites but any discussion of elements of the mission statements is strictly the author's and *does not* pertain to any particular activities these utilities may or may not be doing. Rather this is a discussion of how mission statements can relate to asset management and tie specific elements of asset management together. All subsequent discussion related to these mission statements and the associated activities in the upcoming sections is completely the author's.

4.0 TYING GOALS TO MISSION STATEMENTS

In order to meet the mission statement, the utility will have to take certain actions. These actions can be crafted into level of service goals with specific target levels that can be measured against. As an example, consider the mission statement examples above. Table 2.4 below considers the example mission statement from WaterOne and the potential actions that are required to meet the intended purpose. (As stated above, Table 2.4 is entirely the author's creation.)

Each mission statement will consider different values and desires so the actions required will vary. These actions require resources and support from elected leaders to ensure that the mission can be met. The next step is to take the required actions and create level of service goals around these items. These goals are described in the following section.

5.0 LEVEL OF SERVICE GOALS

Level of service goals should meet the following criteria:

- Meaningful
- Measurable
- Consistent

- Useful
- Unique

Meaningful: The goal should be relevant to individuals who work for or govern the utility as well as any stakeholders. It should be able to demonstrate performance.

TABLE 2.4 Actions Tied to Mission Statement

Mission Statement	Key Words/ Concepts	Examples of Types of Actions Required
"We provide a safe, reliable, high-quality water supply with superior service and value."	Safe	Regulations are met
		No exposure to chemicals (e.g., chlorine) by the public
		No toxic algae or other contaminants
		Lead line replacement program
		Water storage is sufficient to cover fire fighting needs
	Reliable	Replacement of pipe when break rate exceeds a standard
		Source water is adequate to meet the needs of customers
		Water is available 24/7/365 except in times of emergency
	High Quality	Source water is protected
		Regulations are met and potentially exceeded
	Superior Service	Customers are responded to when complaints are made
		Breaks are responded to quickly
		Information is provided to customers in a timely manner when planned disruptions will occur
		Customers are notified to the extent possible when emergency actions are taken
	Value	Rates are set to reflect the service provided
		Rates cover the full cost of operation
		Water is recognized as a vital need for the community

Measurable: It is most important that the goals be written in such a way that they can be measured. It is best if it can be expressed in a quantitative manner, but if not, at least it should be able to be expressed qualitatively and the utility should work towards developing a quantitative way to measure the goal.

Consistent: The goals should be generally consistent with industry practice and measurement of the goal should be reproducible by others.

Useful: The goal should help manage the utility and encourage improvement.

Unique: The goal should describe a specific attribute related to the operation of the utility and each goal should be independent from the other goals. For example, it should not be necessary to use the measurement of one goal to determine if another goal is met.

These attributes are very similar to the criteria that are often cited for "SMART" goals: specific, measurable, attainable, relevant, and time-bound. These attributes can be used in place of those elements cited above if they are more familiar or if they seem more comfortable. Either way, the goal should be compared to the attributes to see if it meets those criteria. If not, the goal should be modified in some way to ensure that it does meet the criteria.

An example of a goal and how it might be compared to the attributes is shown below.

"Address customer complaints related to water quality within 1 business day 90% of the time."

If the SMART goal attributes are used instead, Table 2.6 shows how the goal relates to these attributes.

A question that is often on people's minds is, "how many goals should I have?" The short answer is as many as are needed to properly manage the utility, but that answer is probably not entirely satisfying. Digging a little deeper, it is necessary to have enough goals to document the most important desires/requirements of the customers. In considering the types of things customers want, in general, with water utilities, one would include much of the following:

- *Reliability*—the service is available 24/7/365, whether that is bringing drinking water to the house, taking wastewater from the house, or ensuring that stormwater is collected and handled properly during the vast majority of rain events.
- *Customer Service*—a response is expected when a customer calls with a question, comment, or complaint. Customers also expect to be treated with respect and courtesy. Customers expect to be notified of upcoming, planned disruptions or events that might affect them for an extended period of time.

TABLE 2.5 Relating a Goal With Attributes

Attribute	How the Goal Relates to the Attribute	Comments
Meaningful	The stakeholders are likely to be very interested in a timely response to water quality questions. The staff should also want to respond to these types of complaints. This is a major role of water operators and managers—keep the customers safe.	
Measurable	This goal is able to be measured and it is likely that at least some of the data needed to measure it is available already.	Data needed to measure the goal would include: date the customer complaint was received, the nature of the complaint, the date the utility personnel responded to the complaint. This data could also include the estimated time to repair/replace/address the issue and the estimated cost.
Consistent	This goal is quite likely to be similar to goals at any utility. The specific target of 1 business day may differ from utility to utility, but all should want to respond to customer complaints related to water quality.	
Useful	This goal would be very useful to the utility as it would ensure that managers and elected leaders track whether customer complaints are being addressed in a timely manner. If they are not, the utility management can determine what is needed to improve that process.	
Unique	This goal is written in such a manner that it does not depend on any other goal; it can be measured independently. Further reviews would be required once all goals are written to make sure that it remains unique.	

TABLE 2.6 Relating a Goal With SMART-related Attributes

Attribute	How the Goal Relates to the Attribute	Comments
Specific	This goal is very specific and states what is intended. Every staff member would understand the goal in the same way.	A general goal would be something like, "make sure there is good water." This goal would not be understood the same way by everyone.
Measurable	This goal is able to be measured and it is likely that at least some of the data needed to measure it is available already.	Data needed to measure the goal would include: date the customer complaint was received, the nature of the complaint, the date the utility personnel responded to the complaint.
Attainable	Whether or not the goal is attainable would depend on the level of staffing and how staffing is allocated. The utility should set target levels that are attainable. If 1 business day is not possible, that should not be the target.	The intent with Level of Service goals is to provide the service desired/required by the customers. If customers want a 1 day response time and there are not currently sufficient resources to provide this, the utility should have a conversation with customers. Customers need to understand that additional resources would be required, such as additional staffing and more vehicles, and these needs may require a rate increase to be able to provide the target service level. If customers were unwilling to pay higher rates for the additional resources, the goal's target level would need to be revised accordingly.
Relevant	Similarly to the useful conversation above, it is anticipated that just about any utility would find a goal around customer service response to be relevant to the operation.	
Time Bound	Time is an element that does not apply to all goals, but can be an important consideration. In this case, the time element is 1 day. This metric helps provide guidance for the utility regarding the resources needed to meet the goal.	This element is not explicitly included in the attributes above.

- *Quality*—everyone from regulators to customers to elected leaders expects to have the water treated to required regulatory standards to ensure acceptable water quality.
- *Health*—it is important to protect both public and environmental health.
- *Safety*—it is also important that the safety of the public and employees is protected. Safety may be related to chemicals in use, emergency conditions, or how jobs are completed (e.g., proper barricading, shoring, lock-out/tag-out).
- *Adequate Quantity*—for drinking water systems, customers want to have enough water for their individual needs. In arid areas, this particular desire is one that has to be negotiated carefully so that customers understand the need for conservation and how much water will be available and for what purposes.
- *Appropriate Cost*—customers might state this particular desire in slightly different terms, either "low cost" or "affordable cost" but it should be thought of as appropriate cost. The cost at which the utility will be managed in order to meet all the other aspects that customers want and ensure the long-term sustainability. Similarly to adequate quantity, this is an opportunity to promote the value of water to customers so that they understand the true value of water and all of the complexity of operating a utility.
- *Financial Customer Assistance When Needed*—some customers may face financial hardships for a portion of time or regularly. The utility should consider what type of customer assistance program it may be able to set up to help customers or what other resources might be available to assist customers.

Cathy Bernardino Bailey, Executive Director of Greater Cincinnati Water Works, has stated during conference presentations that when customers cannot afford to pay their bills for basic service, it should be viewed not as *their* problem, but rather "*our*" problem (meaning the utility as a whole). This is an important distinction because it requires the utility to think of creative solutions to the dilemma of household affordability.

- *Adequate Pressure*—customers want to have sufficient pressure in drinking water systems so that water flows from all fixtures, including those in upper floors of houses.
- *Little to No Inconvenience*—repairs, rehabilitation, and replacement can cause customer inconvenience by blocking streets, disrupting businesses, or reducing service for a period of time. While customers may

understand the need for these activities, they may prefer as few disruptions as possible.

If goals can be written around these items to match the requirements of the customers, the utility will be doing well. These goals, associated with customer expectations, are called *external* goals. They address issues that customers and elected leaders would understand and care about. However, it will take many actions, unseen by the public, to meet these external goals. Some of these actions should be written as *internal* goals to enable the utility employees and upper management to track how well the overall operation is working.

As an example, consider a goal related to reliability. "The utility will have no more than 5 percent of its customers without water for more than 5 hours, annually." This level of reliability would require the utility to ensure continuous service throughout the customer base most of the time, which means several things:

- A continuous supply of water
- Few emergency breaks that shut off large portions of the system
- Repair techniques that enable the utility to remain in service during the repair
- Strategic pipe replacement that reduces the number of water breaks per year
- Disinfection equipment that remains in service
- Treatment processes in continuous operation
- Service lines that are replaced when failures are known

A goal with specific metrics could be written around each of these activities to address the internal levels of service needed to meet the external level of service related to reliability. Some examples of these internal goals are shown below:

The utility will maintain a continuous supply of water by ensuring 100% redundancy in water sources.

A valve exercising program will be initiated within 1 year and each valve will be operated at least once per year to ensure proper operation of valves. Any non-operational valves will be replaced within 2 months of discovery

The utility will train its operators on how to use repair techniques that allow the water to remain in service while repairs are being made. When possible, these techniques will be used to make the repair.

The utility will replace 2 miles of pipe annually using strategic pipe replacement techniques to reduce the number of overall breaks by 10% as measured over a 3-year period.

The utility will maintain spare parts for its disinfection equipment to enable the repair of any part within a 30-minute period and prevent water production from having to be shut off any longer than the 30-minute repair time.

The utility will maintain a trained maintenance staff as well as sufficient spare parts to allow the repair of any treatment assets within a 2-hour period. Therefore, the water does not need to be shut off for longer than a 2-hour time frame. This time frame can be covered by water from the storage tank.

The utility will investigate the option to add loop lines to any portion of the system that is not currently looped. If there is a feasible, cost-effective option to add loop lines to the system to reduce the potential for customers to be out of water, the utility will include them in the 10-year capital improvement plan (CIP).

To meet these goals, the utility may need to train staff, purchase equipment, increase staffing, increase the budget, purchase spare parts, install new assets, or any number of other actions. The idea is to consider all of the internal activities that must be completed, monitored, measured, and implemented to meet the external levels of service goals the utility commits to achieve for its customers. These internal goals would likely be measured and reported within the utility staff and management, but would be unlikely to be reported to the general public. The general public, and most likely elected leaders as well, would be most interested in the overarching goal related to reliability, not these internal metrics that make up the actions to achieve the reliability goal.

Coming back to the question of the number of goals, it is important to remember that there is a cost to monitoring, measuring, and reporting on goals, so there is a need to balance the benefits of having goals with the costs. There is also a need to work closely with elected leaders regarding how many external goals will be monitored and how many will be reported publicly, which may be different. In general, six to 12 well-thought-out external goals that fit the most important concerns of the public is a good starting place. Over time, additional goals can be added, if needed, to more fully characterize the utility's operations. The number of internal goals will most likely be higher as several internal goals may be needed to achieve one external goal. A starting point on internal goals may be 15 to 30. Again, goals can be added over time.

As an example, consider the WaterOne mission statement and potential goals that could be set to meet the mission statement. As a start, each element of the mission statement is examined to determine the types of activities it would take to meet that component (Table 2.7). (As a reminder, these are the author's assessment of activities.)

Then consider the goals that can be set to address tone of the activities in Table 2.7—superior service. See Table 2.8 below for some example goals that can be set.

Goals can be set around each of the activities in the mission statement. Additional internal goals can be set for each of these external goals to the extent necessary to achieve the desired service level.

Goals can also be set to address particular issues a utility is having. For example, consider a utility that is having problems with high real water loss (water leaking from distribution and service pipes). Some example goals to address this type of issue are listed below. The target levels are highlighted. These target levels can be adjusted to meet the needs of the utility.

TABLE 2.7 Mission Statement Elements and Associated Activities

Component of the Mission Statement	Activities That Would Help
Safe Water Supply	Routine samples taken and analyzed; flushing completed on schedule; ensuring good supply chain for disinfectants (including backup supplies); chlorine samples
Reliable Water Supply	Reducing breaks/break rate; backup sources; multiple sources; inspections of storage tanks; pump station inspections; pump maintenance
High-Quality Water Supply	Process control sampling; removing contaminants below maximum contaminant levels (MCLs)
Superior Service	Well-trained staff; certified operators; professional staff; training on customer interaction; speed and quality of necessary system repairs; addressing customer concerns/ complaints in timely manner; fully staffed utility
Value	Cost of service is explained to customers; rate structure is fair and equitable; the level of service is equated with the price paid

TABLE 2.8 Goals Related to the Activities Around Superior Service

Activities Related to Superior Service	Potential Goals
Well-Trained Staff, Professionalism and Training on Customer Interaction	All staff receive 20 hours of training per year on topics relevant to their job classification. All staff receive an additional 8 hours of training on customer interaction and professionalism within 6 months of being hired and receive 4 hours of refresher training every year thereafter.
Certified Operators	The system will maintain the required number of certified operators for all of its facilities and operations and the operators in charge will be certified at the appropriate level. There will be pay incentives for operators to achieve higher level certifications and time off will be granted for up to 2 tests per year for certification purposes.
Speed and quality of necessary system repairs	For water: All non-emergency breaks in mains will be fixed within 1 day of the line location process being completed. All break repairs will hold at least 5 years after the repair has been made. If the repair fails sooner than 5 years, an investigation will be undertaken to determine the cause of the early failure. (For a wastewater example: All sewer collapses will be addressed in an emergency fashion [e.g., bypass pumping] within 8 hours of discovery. The sewer will be repaired within 3 months of discovery if it does not require a full pipe replacement [more than 1000 feet] and within 1 year if a full pipe replacement is required).
Addressing customer concerns/complaints in timely manner	Emergency complaints/concerns will be addressed within 4 hours of receipt 95% of the time. Water quality complaints will be addressed within the next business day 95% of the time. Other customer complaints will be addressed within 5 business days 95% of the time.
Fully staffed utility	The utility will maintain staffing at a level such that there is no more than a 5% vacancy rate for staff. When the level exceeds 5% the utility will consider ways in which to increase employee recruitment and retention.

Service lines with more than 1 leak in a 3-year period will be replaced rather than repaired.

Line breaks will be plotted on a map to determine patterns of breakage. Pipe segments with more than 3 breaks will be replaced.

Large water meters will be tested annually for accuracy. A random sample of small meters will be tested every 2 years for accuracy. Any meter that doesn't pass the accuracy test (+/− 3%) will be replaced.

Another example is goals related to safety. If safety is a big concern some potential goals can include:

Every employee will receive 20 hours of safety training per year on topics specifically related to safety issues at this facility.

All incidents resulting in an injury, and any near misses, will be reviewed by the operator in charge as well as the governing body to determine the cause of the injury. Procedures will be put in place to address the cause whenever possible (e.g., purchasing safety equipment, testing safety equipment, improving lock out/tag out procedures, etc.).

The most important aspect of Level of Service goals is to ensure they help the utility deliver the service desired by the customers and the elected leaders. They should be structured around the most important factors of the mission statement and the most important concerns of the utility.

6.0 WHEN THE GOAL IS NOT MET

One reason to have Level of Service goals that are sanctioned or approved by the elected leaders/governing body/owners or ultimate decision-makers is to enable the conversation around the actions that are required when the goals are not met. If this happens just once or twice, the response may be to do nothing and wait to see if the situation improves. If it happens regularly, and it is clear the situation is not going to improve without some type of proactive intervention on the part of the utility, it is time for a serious discussion of what needs to change to improve the situation.

Some required actions can be relatively easy. If the goal is simply beyond the ability of the utility to meet and there is little effect on customers, the target level for that goal can be adjusted or the underlying goal itself can be revised.

Example: The goal is: We will provide adequate and acceptable water pressure throughout the system, with a target average pressure of 70 psi throughout the system 95% of the time and a target minimum pressure of 40 psi. (Note: Within this goal, the target levels are: 70 psi, 95%, and 40 psi. Utilities can adopt similar goals to each other but set different target levels, depending on their customer requirements, regulatory climate, and system capabilities.)

Pressure measurement by the utility reveals that the average pressure is closer to 60 psi and that there are portions of the community that experience frequent pressures of 35 psi. It is determined by the engineers through a hydraulic model that the utility will function properly at these lower pressures. The state only requires 30 psi as a minimum so a pressure of 35 psi will still meet the state regulations. The utility seeks feedback from business customers and none of them indicate the lower pressure is a problem. The utility determines that residential customers will not complain about a pressure of 60 psi versus 70 psi because many of them are already receiving 60 psi water and are not complaining. A check of customer complaint logs backs up the claim that customers do not have a problem with the lower pressure. The decision is made to adjust the target levels and monitor whether or not this pressure can be met and whether there are any negative effects to the water utility operation as a whole.

The purpose of Level of Service is not to create goals that are difficult to meet without a lot of extraordinary effort; rather the purpose is to communicate what the utility is going to provide the customers and how well it is doing that. Therefore, service level statements should not be set much higher than what is feasible. This is not a situation in which the acceptable is considered a grade of "C" and the utility should be striving for a grade of "A." If the utility is providing the Level of Service it has promised its customers and that level meets customer expectations, that is the equivalent of an A. The target levels should reflect what is achievable and within the range of what is desired by the customers.

In the first example, the utility only needed to change its target level, not the underlying service level statement about providing adequate and acceptable water pressure. In other cases, more effort may be required to address the non-attainment of a goal. In these cases, it is important to address the root cause(s) of why the goal is not being met. Consider the following example:

The wastewater utility will protect the environment through appropriate collection system practices, with a target level of no more than 20 sewer backups related to fats, oils, and grease buildup in a year. (Note: Within this goal, the target level is: 20 sewer backups.)

If the utility experiences 25 or 30 backups, it must consider why this occurred. Maybe the root cause was one area of the community that has a large number of restaurants. These restaurants are supposed to install and maintain grease traps but the utility has never checked to see whether they are complying with the regulation. The first action the utility is going to undertake is a random inspection program to determine if the restaurants are complying with the grease trap requirements. They will be checked to ensure they are in place, are the correct size and type, and are being maintained properly. Any restaurant found in violation will face a fine of $200 per infraction. Perhaps in addition to this type of action, the utility needs to increase cleaning in some parts of the collection system or some pieces of pipe need to be replaced. Meeting this goal may require an investment of funds, such as personnel to do inspections or additional sewer cleanings, because it is quite possible that the customers will not agree with raising the target number of backups to a higher number.

These types of actions may not be necessary the first time a target level is missed. The goal should be monitored for a period of time to see if the situation corrects itself through the normal course of business. If that does not occur, more significant action can be undertaken. Also, it may take a longer period of time to have valid results of monitoring. If there are very few occurrences in the specified time period (e.g., month, quarter), a few incidents could cause the entire goal to be missed. If there are many incidents occurring, a more valid analysis can be done.

There are some goals, however, for which not meeting the goal does require immediate action, such as goals related to regulatory requirements. If these goals are not met, it is important to understand why not and to act quickly to rectify the situation.

7.0 WHERE TO SET TARGET LEVELS

For all non-regulatory goals or those that are not set by the elected leaders, the target levels can be chosen by the utility staff itself. There are two boundary conditions for goals and their associated target levels: it is not acceptable to go below the regulatory requirements and it is not possible to go beyond the physical capabilities of the assets. For example, a pump can only pump as much water as what it is rated for, a well can only produce so much water, a storage tank only holds a certain number of gallons, and these assets cannot be made to achieve levels outside their capabilities. The regulations at the local, state, and federal level are requirements that must be met at all times. The utility should never expect to operate on a routine basis that does not meet these levels. If the utility finds itself in non-compliance, this should be considered unacceptable and an indication of a goal not met. An appropriate action should be taken to prevent future non-attainment.

Between the boundary conditions there is considerable room for the utility to set goals and associated target levels based on what the customers want and what the resources will provide. It is highly likely and expected that utilities will set different target levels even for the same goals. Consider a goal around average chlorine residuals and minimum residuals. One utility may set this level at 0.5 mg/L average and 0.2 mg/L minimum while another will set the level at 0.8 mg/L and 0.4 mg/L, respectively. Each utility should carefully consider its needs and resources and set the target levels accordingly. It is also possible for the target levels to be adjusted over time as the utility improves its asset management practice or as conditions change.

For example, consider a goal around sewer backups. Perhaps the utility currently has 60 sewer backups per year on average and would like to eventually go to less than 30 per year on average. It may be very difficult to take all the actions necessary to reduce the number from 60 to 30 in a short period of time (a year or two). Therefore, the utility may wish to start with a target level of 50 backups then reduce it by five to 10 backups every year or two until the goal of 30 is met. Or it could choose to reduce to 20 backups over a 5-year period then reduce another 10 backups over the next 5 years until it reaches five backups or fewer per year. Part of the reason it is possible to adjust target levels is that level of service goals have bands around them. This concept is explained in the next section.

8.0 GOALS HAVE SERVICE LEVEL BANDS

The target levels for goals contain service level bands, meaning that customers may find it acceptable for the utility to operate anywhere within the band. Above or below the band may be too costly for the utility, outside the range of the physical capabilities of the assets, or may be unacceptable service. Some of these bands may be quite large, while others are very narrow. There are a few goals that have no service level bands and these goals relate to utility requirements, such as regulations. In these cases, there is the potential to go beyond the goal to remove more of a contaminant, but there is no option to remove less of a contaminant.

It is important to consider the balance between customer desires and costs for individual goals and how that relates to the bands around the goals. For some goals, adjusting the target level to increase the service provided to customers will have a minor effect on cost while in other cases it may have a major effect on cost. For example, consider a goal around speed of repair for a water main break. Suppose the goal is "non-emergency water main breaks will be fixed within 3 days of discovery 95% of the time." Because water main breaks typically require the response of a one-call type service where other utilities need to be marked prior to excavation of the water line, there will be a limit on how fast the pipe can be repaired. In any given state it may take 1 day to a week to have the one-call service respond. Therefore, the lower end of the band is set by the speed of that service; the utility does not want to promise a faster service than they can deliver, especially since this part of the speed is outside their control.

On the other end of the band, customers may accept anywhere from 3 days to 2 weeks, depending on how arid the climate and how aware customers are of conservation. In more arid climates, the break repair would be expected to occur more quickly, while in in more water-rich areas, the customers may tolerate a longer repair time. The time chosen may have some effect on cost if breaks tend to happen in clusters. If, for example, a few days of cold weather causes several breaks to occur at once, it would be very difficult to fix them all within 3 days without paying overtime or having additional crews or additional vehicles. Any of these actions would raise the cost of meeting the goal. The balance needs to be struck between how important it is to respond quickly and fix the breaks versus the costs that will be incurred to do so. If this type of service is very important to customers and they would be willing to pay more or would prefer the utility to put its resources towards this activity versus another one, then keeping a short time frame would be reasonable. If, however, customers are not that concerned about break repair, and water is not scarce or expensive to produce, then the extra expense may not be justified and it is better to set the target level at a higher number.

The idea of service level bands described by Ross Waugh, a New Zealand Asset Management expert who is also one of the contributing authors of the International Infrastructure Management Manual (IIMM) and the author of Inframanage.com (https://inframanage.com/), an asset management website, is portrayed below. Each of the colored boxes represents a different type of goal category, such as customer response time, regulatory requirements, pressure, pipe breaks, sewer overflows, flooding occurrences, or any other category. The x-axis is not a spatial representation. It is just meant to convey that there are different types of goals. The y-axis is a spatial representation of how much opportunity there may be to change the target level. If the height of the box is small, there is little opportunity. If the height is large, there is much more opportunity.

The magnitude of the service level bands depends on several factors, such as cost implications, resource availability, community demographics, customer desires, regulatory flexibility, capability of the assets, and utility

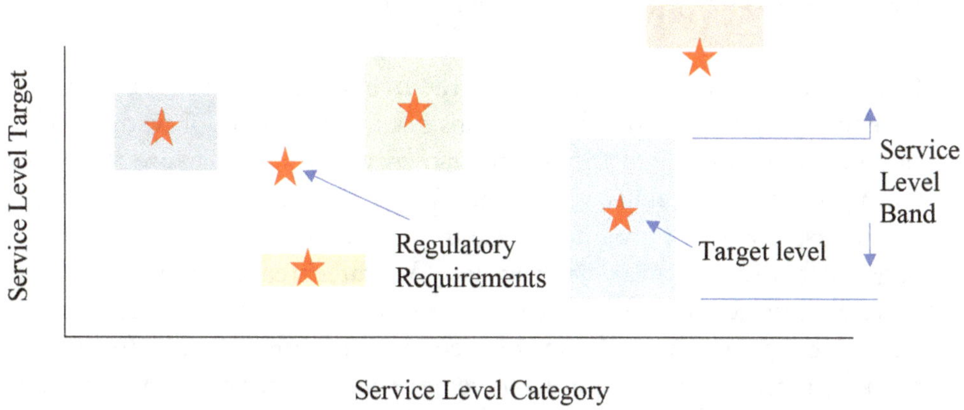

FIGURE 2.1 Service Level Bands

managers' and elected leader's considerations. Note the variation between level of service bands between categories and imagine the types of goals that would have smaller or larger bands. Different utilities would have larger or smaller bands for the same types of goals, depending on their particular customer expectations. Given these variations, it is important to ask customers about their concerns and desired level of service to ensure that the utility is matching customer desires with the goals that it sets. This involvement is discussed more in the next section.

9.0 CUSTOMER SERVICE INPUT

The best way to determine the desired level of service is to ask the customers. There are several ways to accomplish this task, including: surveys by phone, mail, or electronic means; focus groups; public meetings; annual meetings; or any other method that is likely to reach the utility's customers. It might be difficult, initially, to gain this input so it is possible to begin by setting the level of service goals based on assumptions about customer needs and then gathering the input over time. Then the goals can be adjusted as needed based on the responses.

It is best to remember that if customer input has not been sought, the service level targets may or may not accurately reflect the customer desires. Therefore, the utility may set target levels that are higher or lower than what customers want. As input is sought it is important to consider all voices, not just the loudest voices, and to make sure the voices cover a range of customers across the utility.

10.0 GOALS AND COSTS RELATE

Goals and costs are interrelated. As the overall level of service increases over time, the costs also increase. Consider the beginning of water service. Initially, individuals had to haul their own water and, because the water was untreated, the quality was poor. Wastewater service was an outhouse, where waste would flow into a hole in the ground. This service level was very low but the cost was also very low. As water treatment increased over time, the service level also increased but so did the costs. Now that wastewater treatment is a minimum of secondary treatment and, often times, advanced treatment is required to remove nutrients, and drinking water must remove contaminants to meet the National Primary Drinking Water Standards (NPDWS), the costs have gone up considerably, but the level of service is quite high. While at the turn of the 20th century, water-borne illness was the third leading cause of death in the United States; by 1999, water-borne illness did not make the top 10 causes of death list (Centers for Disease Control, 1999).

As major increases in service levels occur, costs can rise significantly. However, as the industry responds to these requirements, costs usually level off and may even decrease slightly. For example, when wastewater treatment went from primary treatment to secondary treatment, a significant investment had to be made to reach this new plateau. Once reached, technological advances reduced any further increases related to that specific change and costs leveled off. As treatment advanced to tertiary treatment, costs again increased. But there are many more technologies available now to achieve tertiary treatment, limiting further cost spikes related to this service level change. It is very important for customers to understand that each of these increases in service levels provides public health, environmental, and social benefits that make the investment in the increased service worthwhile.

While the overall level of service clearly relates to cost, the relationship between target levels and costs for individual goals is more complex. For some goals there is a more direct relationship in which the higher the target level, the higher the cost. For other goals, there may be a very small increase in cost for a large increase in the target and for still other goals, increasing service levels a small amount may cause an extremely high cost increase. It is important to understand the relationship between costs and service so that the potential effects can be communicated both within and outside the organization. This is particularly true for individuals who are making decisions on where to set target levels. Choosing to increase service in one area may reduce funding to address another area.

As an example, consider the goal of sewer backups in the city over the course of the year and assume the utility decided to set that level at zero backups. Suppose the average number of backups per year as currently measured over the past 10 years is 30. Reducing the number to zero is going to take additional effort that is not happening now that might include: additional sewer cleaning; repair of pipe segments; replacement of a pipe or pipe segments; or additional pretreatment or work with businesses to reduce inputs of fats, oils, and grease. It might also include the need for additional staff or equipment. If the system wants to set the goal to zero backups, it needs to be aware of these potential costs and whether it is willing to invest the money to reach this goal. It should also consider the opportunity cost of the investment in this goal. If the money is invested in this goal, it will not be available for investing in other goals or other activities. A decision needs to be made regarding whether the goal is realistic for the system and, if not, a goal that more closely matches the ability of the system needs to be considered.

Another consideration is what the customers are willing to accept in terms of backups. If the system's customers feel that no backups can be tolerated, the target level for that goal can be set there but it is then necessary to communicate with customers the potential rate implications. If this goal means that the rates may need to rise by some amount, this can be communicated with customers. If they support additional funds being spent in this manner, they should be willing to support rate increases. If instead customers push back and expect the utility to "do more with less" or find "free money" to reduce costs, the utility should explain that the reasons why these approaches will not work (if indeed that is the case). Public communication around these issues is crucial to ensure there is public support.

As another example, consider a system who has one or two water quality complaints each month, with a goal to respond to water quality complaints within 48 hours. If they wish to change the goal to within 24 hours, there may be limited cost effects because there are few complaints and the change in response time is not that significant.

Choosing goals and setting target levels has an effect on cost, which can be minor, moderate, or significant and these effects should be considered as well as the need to communicate within the organization to utility managers or executives, and with elected leaders, and customers regarding these effects.

11.0 COMMUNICATING GOALS

There are different audiences in which to communicate the goals and the methods of communication vary depending on who is receiving the information. The audience and potential communication methods are shown on Table 2.9 below.

TABLE 2.9 Communicating Goals

Potential Audience	Methods of Communication	Comments
Water System Personnel	Regular staff meetings	Specific Level of Service goals may be communicated during regularly scheduled staff meetings. These meetings may cover all or some of the Level of Service goals and progress towards meeting the goals.
	Internal Electronic Sharing Platforms	Some utilities have used technology that allows individuals to access a shared site that allows for communication or access to stored documents. The Level of Service goals as well as a document indicating current progress towards meeting the goals can be stored in the shared document file system.
	Postings on bulletin boards	If there is a bulletin board or other central place in which information for employees is posted, the Level of Service goals can be posted there along with a current listing of progress towards meeting the goals.
	Dashboards	There are technologies now that can access stored information on spreadsheets or databases and present a dashboard of the results. It is possible to include Level of Service information in the dashboard so that employees can examine the status of meeting the goals at any time.
Upper Management	Dashboards	See above.
	Internal Electronic Sharing Platforms	See above.
	Preparation of Level of Service Reports	Upper management is likely to contain the employees who prepare the Level of Service information. The act of gathering the information and preparing reports will keep upper management informed.
Elected Leaders/ Board Members/ Governing Body	Regular public meetings	Most elected leaders/governing bodies have regular meetings (e.g., monthly, quarterly). Progress towards meeting the Level of Service goals can be presented at each meeting or at meetings on regular intervals (e.g., every other meeting, once quarterly).
	Annual Summary	An annual summary can be prepared that describes the progress towards goals for the entire year. The report can be delivered electronically or in person at an annual meeting.

(*continues*)

TABLE 2.9 Communicating Goals (*Continued*)

Potential Audience	Methods of Communication	Comments
	Dashboards	See above. The elected leaders could be granted access to dashboards.
	Internal Electronic Sharing Platforms	See above. The elected leaders could be granted access to the sharing platforms.
Customers/General Public	Consumer Confidence Report (CCR)	The CCR has required elements but it is allowable to include other pertinent information, such as ability to meet Level of Service goals in the CCR. This report is an annual requirement that is made available to all customers, so it represents a good opportunity for customer education.
	Billing Inserts	Systems routinely send bills to customers. The bills may have places on them for additional information, such as Level of Service goals or an additional page can be added that would include this information. Most likely, this would be once a year as opposed to every bill.
	Website announcements	Many utilities have websites that provide information to the public. The Level of Service information can be placed on the website.
	Presentation at an annual customer meeting	If the water system has any type of annual public meeting, the Level of Service goals as well as progress towards meeting these goals can be included on the agenda for the meeting.
	Presentations at open, public meeting	Most publicly owned water systems have routine public meetings, such as monthly, or quarterly. The public is entitled to attend these meetings and meeting minutes are prepared that are also public. If Level of Service goals as well as the ability to meet these goals is part of the regular meeting, the public will have access to this information.

When communicating goals to elected leaders and the general public, it is generally helpful to keep the communication short, to the point, and easily understood. It should easily and quickly communicate the main points: which goals are met, which are not, and which are close to the goal (either just above or below the goal target level). It is important that the leader can easily zero in on the key issues and quickly focus discussions on the most important items.

One example of communicating goals is shown below on Table 2.10.

TABLE 2.10 Level of Service Goal Communication Tool

Category	Goal	Target Level	Actual Performance	Status
Customer Service	The water quality complaint rate per 1000 customers will be less than 1.	Less than 1 complaint per 1000 customers	0.9 complaints per customer	▲
Maintenance	The ratio of planned maintenance to corrective maintenance will be 70%/30% based on the number of work orders processed.	70/30	67/33	■
Training	Every employee will receive 20 hours of training each year.	20 hours/ employee	18	■
Collection Efficiency	There will be fewer than 20 sewer backups per year due to wet weather conditions less than 100-year storm events.	20 backups	15	▲
Water System Efficiency	The utility will perform a water audit annually using industry standard methods to determine non-revenue water and will report the results annually to the public.	Annual audit and reporting	Annual audit completed. Reporting not done.	▼

▲ Target Met ■ Target Not Met but close to target level ▼ Target Not Met

A simple tool such as the one shown on the table can alert upper management or elected officials very quickly to the status of level of service goal performance. Reviewing such a chart would only take a few minutes to determine which areas were performing well and which were concerns. Using a chart like this, an elected leader could easily focus discussions on the last goal—water system efficiency which is not currently being met.

If it was an infrequent occurrence for one of the goals to be missed, a simple discussion to determine the reason the goal was missed could be had. If it was caused by an event or situation that could be remedied easily or by something not expected to happen again, the best course of action would probably be to reevaluate at the next reporting period and not take further action. If instead, the lack of meeting the goal was due to something that would require further investigation, a dialogue could be held to discuss the issues/concerns and potential actions that could be taken to address the situation. If this was a situation in which the goal had not been met for several reporting periods, the board could discuss appropriate actions to take to improve the ability to meet the target levels of the goals.

For goals that were not met but were close to the target, it might be best to monitor for a bit longer to see if the goal is met at a later date. If the goal remains just below target level for a period of time, similar to the discussion above, a dialogue should begin regarding what actions can be taken to address the situation and improve performance.

In either case, if actions are taken and it is still difficult or impossible to meet the goal, a discussion can be had regarding the target level and whether the target level should remain where it is or be adjusted up or down as the case may be to a level that can be met. This type of action should not be taken lightly and should not be done solely because the goal was missed. The discussion should include several factors. Some factors will make it difficult to revise a target level such as:

- Is this a goal that must be set at the current target level (i.e., a regulatory or otherwise mandatory goal)?
- Is this a goal that meets an internal requirement such as a board directive, a municipal ordinance, a customer agreement?
- Is the goal one that customers have strong opinions about and would object to changing the target level?

As an example, a utility in Australia had a goal related to the speed with which water main breaks were repaired. They believed their customers wanted breaks to be repaired within 6 hours of discovery. They were working very hard to meet this goal but as the break rate increased, were finding it difficult. The choice they faced was whether to continue to meet this goal, which would require additional staff and equipment, potentially increased costs and perhaps a rate increase, or to revise the goal to take a bit longer to repair the pipes. When they reached out to their customers, they determined that the customers were not that concerned with having mains fixed within 6 hours and would accept 24 to 48 hours. However, what they really wanted was to have some notice that the water was going to be shut off and repairs were going to be made. The customers were actually unhappy with the speed of the repairs because it meant that service was shut off quickly and without notice. Therefore, the utility was able to change its goals to one of fixing the breaks within 24 hours but customers would be notified at least 15 minutes before shut off. This approach improved customer service and saved the utility considerable dollars.

Sometimes a change in the level of service target level is the best way to address the issues.

Other factors will make it easier to make a change, such as:

- There is a large band around the goal level (see Figure 2.1) so moving the level will cause little concern for customers.
- The goal is primarily an internal rather than an external one.
- The change will not result in any increase in health and safety concerns for customers or employees.
- The goal is not tied to any regulatory issues or internal requirements.

If it makes sense to change the goal rather than continuing to try to meet the goal and it meets the criteria for a goal that can be changed, that route can be taken and in some cases will be the preferrable option.

12.0 MEASURING GOALS

Goals can be measured daily, weekly, monthly, quarterly, semi-annually, or annually. The frequency is likely to be different for each goal; however, all goals should be reviewed at least annually. The frequency of measuring an individual Level of Service goal depends on several factors including:

- The amount of data available
- The likelihood of change within the time frame
- The importance of acting on the information in a timely fashion

Some goals do not change frequently or there is insufficient data between time frames to measure a change. In other cases, it is not necessary to monitor the information that frequently. Some examples of potential types of goals along with a potential frequency are shown in Table 2.11.

One additional consideration is the cost of monitoring and measuring level of service goals and associated targets. In some cases, the utility collects information already that can be used to measure the target levels (e.g., regulatory compliance samples and analysis) that minimizes costs. In other cases, the utility may not yet be collecting data and it must put practices in place to start obtaining the data. These costs may involve personnel time, monitoring equipment, technology purchases, or other related expenses. The costs may affect how often a goal is measured or if the goal measurement must be phased in over time to allow for the purchase of necessary equipment. The utility should plan and budget for the costs required to ensure that the goals can be measured at the frequency desired.

TABLE 2.11 Level of Service Goals and Frequency of Measuring

Type of Goal	Frequency	Comments
Regulatory	Daily, weekly, monthly, quarterly, annually, less often than annually	The frequency of regulatory goals depends on how often the regulations require data collection. For example, a surface water plant may collect filter efficiency data throughout the day, so checking this information for regulatory compliance may need to be daily or weekly. Drinking water coliform data is collected monthly so analyzing this data may be monthly or quarterly. Data for some contaminants, such as nitrates, may be annually so there is no need to monitor that more often. National Pollutant Discharge Elimination System (NPDES) permit data is often collected daily with a required monthly report. This data should be reviewed daily, weekly, or monthly for compliance with regulations. This data is also more frequent because the need to act is urgent. If a regulatory goal is not being met, the system will need to take immediate action to improve the situation. Some regulatory tests are taken less often than annually so they do not require annual review of goals.
Customer Service	Monthly, quarterly, semi-annual, or annual	If there are frequent requests for customer service from the utility, it warrants checking more frequently. However, if there are few complaints or requests for service, semi-annual or annual are frequent enough.
Asset Performance	Quarterly, semi-annual, annual	Goals related to asset performance, such as number of sewer backups or pipe breaks, may take a longer period of time to see trends. The number of expected occurrences will indicate how often is reasonable. Additionally, once a robust preventative maintenance program is established, the schedule of maintenance activities may dictate the frequency of when goals are measured.

(*continues*)

TABLE 2.11 Level of Service Goals and Frequency of Measuring (*Continued*)

Type of Goal	Frequency	Comments
Water Efficiency	Annually	Water efficiency in drinking water is measured through water loss audits, while in wastewater it is measured through infiltration/inflow (I/I) records. Generally, water audits are done annually, and the annual records can be compared from year to year. I/I can also be viewed annually and compared from year to year to see trends. Efficiency trends can be harder to see on a monthly basis as many other factors need to be considered when a shorter time is used.
Utility Management	Annual	These goals relate to the overall management of the utility and how it is implementing programs like asset management. It may be hard to see any changes on a short-term basis as many of these activities take time to implement, so annual is probably a good time horizon.
Internal Utility Goals	Monthly, quarterly, semi-annual, or annual	These goals may vary in terms of the best timing as some will require more frequent review and others less frequent. Each internal goal should be examined to determine the most appropriate time frame.

13.0 REFERENCES

Atlantic City Municipal Utilities Authority. (n.d.). *Mission statement*. https://acmua.org/mission-statement/

Centers for Disease Control. (1999, July 30). Achievements in public health, 1900–1999: Control of infectious diseases. *Morbidity and Mortality Weekly Report*. *48*(29), 621–629. https://www.cdc.gov/mmwr/preview/mmwrhtml/mm4829a1.htm#fig1

Central Arkansas Water. (n.d.). *Mission, vision, and values*. https://carkw.com/about/mission/

Mojave Water Agency. (n.d.). *Strategic plan*. https://www.mojavewater.org/about-mwa/strategic-plan/#:~:text=Mission,risks%20proactively%20using%20sound%20science

Verona. (n.d.). *Stormwater utility.* https://www.veronawi.gov/485/Storm-Water-Utility

WaterOne. (n.d.). *Mission, vision & values.* https://www.waterone.org/341/Mission-Vision-Values

Chapter 3

Asset Inventory

1.0 WHAT IS AN ASSET?

If you are going to create an asset inventory, one of the most fundamental questions is, what is an asset? A simple definition is: anything you own that has value. However, this definition is not going to be sufficient. The first part can be problematic because there may be assets that are not owned, but are managed. It is important to include these assets in the inventory if one is going to be performing operation and maintenance tasks on them or routine inspections. Green assets (which will be discussed in Chapter 7) are a good example of assets that may not be owned by the entity managing them. Other assets that may fall into this category are those that are owned by one governmental entity but managed by another. For example, a county might own an asset that a city within the county manages or vice versa. Another example may be stormwater assets that are privately owned but managed by the utility or that are owned by one governmental entity but operated and managed by a utility that is not owned by the same governmental entity. Therefore, the definition can be expanded to anything you own or manage that has value. This expanded definition addresses the first problem with the definition but does not address the second problem. How much value?

Not everything within the utility has sufficient value to be worth managing as an asset. In this case "value" is not just a monetary idea but is also an operational idea. What is the overall value of the asset within the context of the utility? How much value does it have to the overall purpose of the utility? Now we need to change the definition to anything you own or manage that has sufficient value to be tracked, tested, and assessed. Initially, this idea that not everything in a facility is an asset may seem like a strange concept. The item would not be in the facility if it did not have a purpose, right? If it is there, it must be important. If we took it out, the facility would not function properly, so everything has to be an asset, doesn't it?

It is important to look at these items within the facility another way. How do you *manage* each item? There are some things that require daily checks, frequent maintenance, or routine repair and there are other items that may not receive any attention at all. Consider a few examples.

A filter on a wastewater treatment plant blower. This item is very necessary because it filters the intake air and keeps impurities out of the blower. Its presence prolongs the life of the blower. It has *value* to the utility, but is it an asset? Let us consider how this filter is actually managed. When the filter is observed to be clogged or contains excessive debris, it is removed and thrown away and a new filter is put in its place. Therefore, this filter is not *managed.* It is just a necessary supply or consumable and should be treated as such. It is not an asset and does not warrant treatment as if it were. Consider if it was treated as an asset. We would need to collect information on the filter, give it an ID number and track when it was replaced. This would be a lot of work for very little to no value. It is possible to have items, such as the filter described here, that are not assets, but may be tied to an asset. For example, if the utility wished, it could tie the filter component in a spare parts inventory or in a "supplies" list to the blower asset it is associated with.

Now consider a blower at a wastewater treatment plant. The filter was not considered an asset, but is the blower itself? The blower is needed to supply air to the activated sludge basins and without the air, the treatment plant will not function. The utility would violate permit limits and potentially pollute receiving waters. The blower performs a very important function so it has value to the utility. The next question is how is this asset managed? This asset would receive routine operation and maintenance and would be repaired when it was facing a failure. Information about the blower is valuable to the utility, especially if there are multiple blowers and the utility wishes to track trends with the assets (time between maintenance activities, types of repairs, frequency of repairs, trends between frequency of repairs and failures, cost of operation, energy use, and operational run time). The blower is an asset because it has both value and it is worth managing as an asset.

These two examples may seem relatively easy to identify. It can get a bit trickier to determine in other cases, so setting some guidelines can help make the decisions regarding what is and is not an asset a bit more objective. Potential considerations include:

- Assets must be above a monetary amount in purchase cost. This value is typically higher for large utilities and lower in smaller utilities. For example, a large utility might set the cost at $3000 to $5000 dollars while a small utility might choose $500 to $1500. The idea is to balance the dollar amount so that neither too many nor too few assets are in the inventory. Too many assets renders management of the asset inventory too difficult and makes it hard to sustain the information and quality over time. Too few leaves out so many assets that insufficient information is known about the items that make up the system. It may be necessary to adjust the dollar amount over time to achieve a reasonable balance.
- Assets in which the cost of a single asset is below the threshold chosen above but for which there are a large number of assets in the overall system should be included. In this case, an individual asset, such as a water meter or a fire hydrant, may be below the threshold value but because there are so many in the system, collectively the set of assets is well above the threshold. Furthermore, these types of assets are generally managed. In this case, these items should be considered assets because leaving them out would make it very difficult to effectively and efficiently manage the system.
- Anything that a work order is written on and data is collected as part of the work order should be included as an asset. This consideration does not include a simple work order to change out a supply item. For example, if a work order was written to replace the blower filter, no data other than the date the filter was changed would be collected and this type of work order would not make the filter an asset. In the case of a supply item, like a filter, a better practice would be to write the work order on the parent asset—the blower—rather than writing

a work order on the filter. Considering how work orders are written also helps identify how many assets make up a parent asset. This idea will be discussed in the next section on Asset Hierarchy.

- Highly critical items for which a failure of the item would cause a substantial problem for the facility need to be considered assets. There may be items that are highly critical to the operation, but they may be below the threshold and perhaps are not managed as assets. However, their failure may cause a cascade of problems that would negatively affect the utility. In these cases, the items should be included as an asset. An example may be sensors that control the air in an oxic/anoxic treatment unit. If the sensor is not reading properly, insufficient or excessive air may be added to the treatment tank causing a disruption of the treatment process. These items can be considered assets to ensure they are routinely checked and their operational data tracked to assess how often the asset should be replaced.

These criteria can be used in combination with each other to ensure that the correct set of assets is included in the inventory. Additional criteria can be established to address particular needs of an individual utility. It may be helpful to develop a logic flow chart to ensure that there is consistency in establishing assets across the organization. It is important that each part of the utility treats assets similarly. Figure 3.1 provides an example.

2.0 ASSET HIERARCHY

The asset inventory should be created as an asset hierarchy meaning there are parent, children, grandchildren, and great-grandchildren assets. The parent is the main asset with children assets associated with the parent and grandchildren assets associated with the child asset. The hierarchy allows the utility to see connections between assets and roll-up costs to see costs associated with various portions of the utility. For example, a pump station building can be the parent asset. The children assets would be the wet well, the pumps, a crane, piping, and HVAC. The pump can further break down into the motor and the controls. The piping children assets can include the meter, valves, and the pressure gage. Therefore, costs can be rolled up for each individual pump, the entire pump station, or other combinations to see where and how money is spent on assets.

A simple example of a wastewater hierarchy is shown in Figure 3.2. The parent asset in this example is the Wastewater Treatment Plant so all assets below can be rolled up to get the costs associated with the treatment plant itself. The next level—children assets—includes: Headworks, Primary

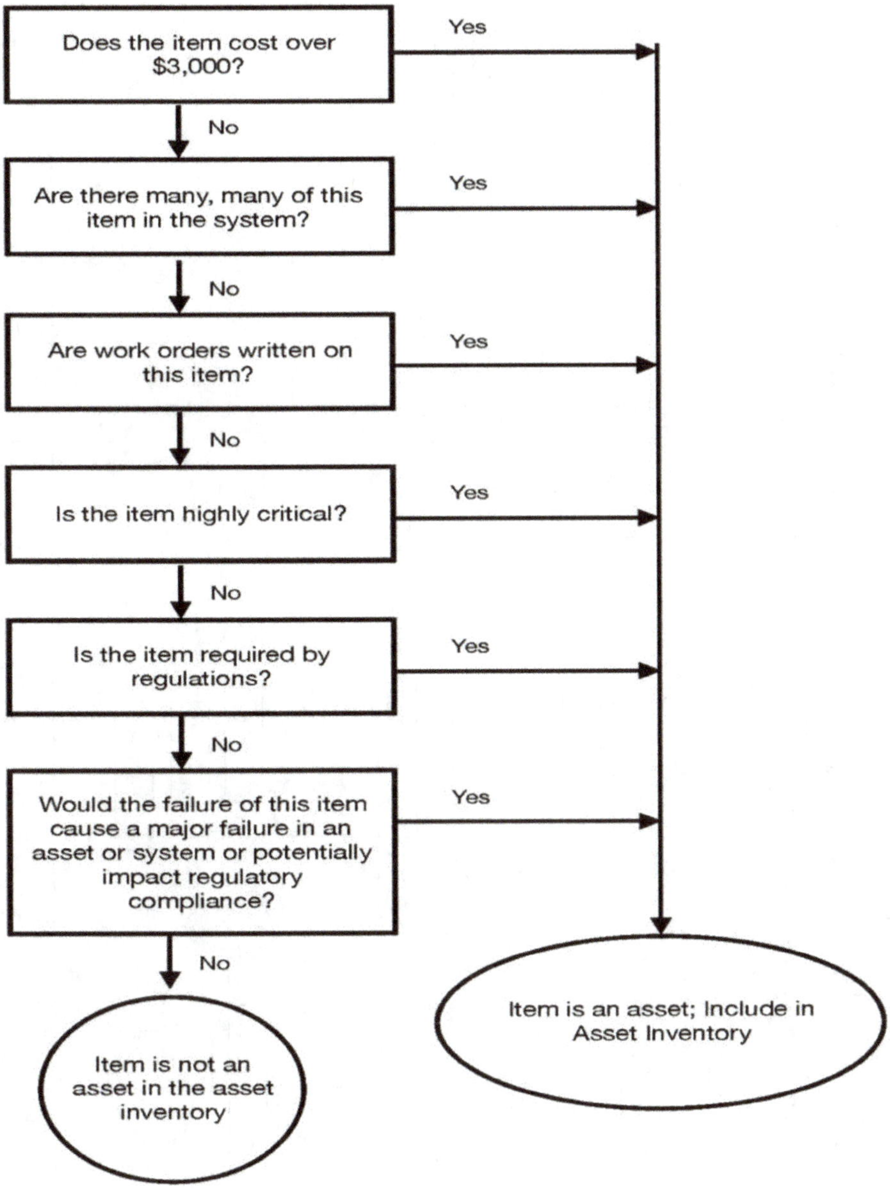

FIGURE 3.1 Asset Identification Flow Chart

Treatment, Secondary Treatment, Disinfection, and Solids Handling. Therefore, assets, costs, and other information can also be rolled up in these categories. The next levels of assets include grandchildren, great-grandchildren, and great-great-grandchildren. Clearly, this is a simplified hierarchy and the number of assets would be far greater than what is shown. It is provided an illustration of what can be done and how assets can be listed in a connected way to permit comparisons of work orders, activities, costs, numbers of

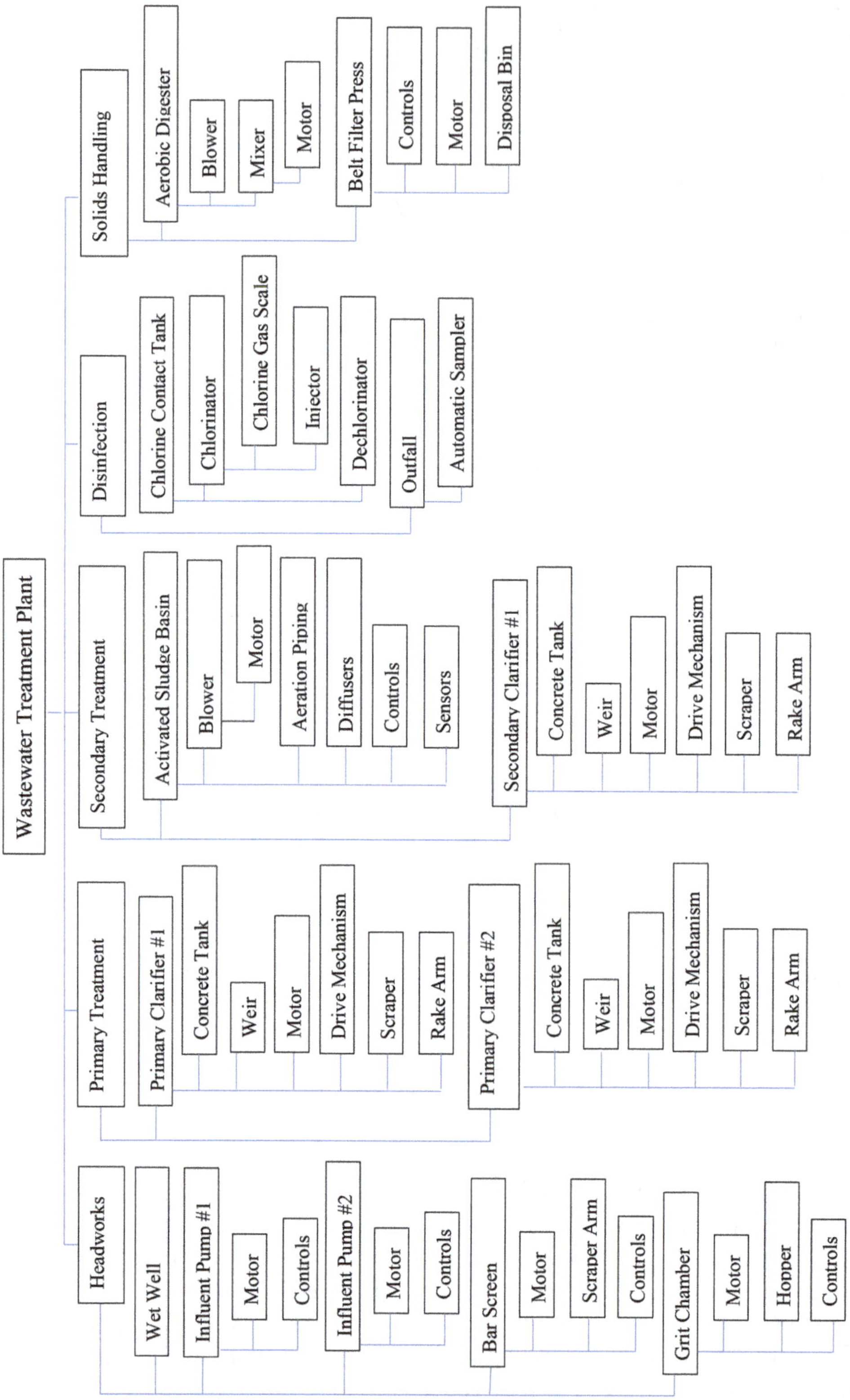

FIGURE 3.2 Asset Hierarchy Example

assets, criticality, and many other comparisons. If assets are not organized in this manner, it is much more difficult to make comparisons of where resources are being spent. That information provides an opportunity to determine if too many resources are being spent in one part of the facility. On Figure 3.3, a portion of the hierarchy is shown so that the relationships between the assets (parent, children, grandchildren) can be demonstrated a bit more clearly.

The figures represent a process-based hierarchy. There can also be location-based hierarchies, meaning a hierarchy built around physical locations. There can also be a combination of the two types in which a location-based hierarchy is used, but within each remote location, the assets are listed in a process-based hierarchical way. The operations and maintenance staff can determine which methodology will work best to enable them to tie assets together in a meaningful way.

Similarly, the collection or distribution system assets can be organized in a hierarchical way although sometimes this is done through mapping software rather than asset management software. There can also be hierarchies for specific distribution or collection assets, such as booster pump stations or sewer lift stations. Separating assets by pump stations can permit an examination of costs by station that can be illuminating to the managers of utilities.

3.0 ASSET ATTRIBUTE INFORMATION

Every asset has important information that is associated with it, which is often called "attributes." These attributes describe the asset in various ways and having this information readily accessible is incredibly valuable for operators and managers. There are a few types of attributes that are necessary to collect on each asset, but it is possible to collect additional attribute information to best meet the needs of utility personnel. While there are quite a large number of attributes that can be obtained, it is important to only collect the attributes that will be valuable and useful to the organization. Collecting additional information beyond those useful attributes will take time and resources but will produce little to no benefit afterwards. Therefore, the benefit of the attribute information should be balanced against the cost and personnel effort to collect and manage it.

The attributes that should be collected, at a minimum, on each asset include the following:

- Asset ID (unique identifier for each asset)
- Asset Name

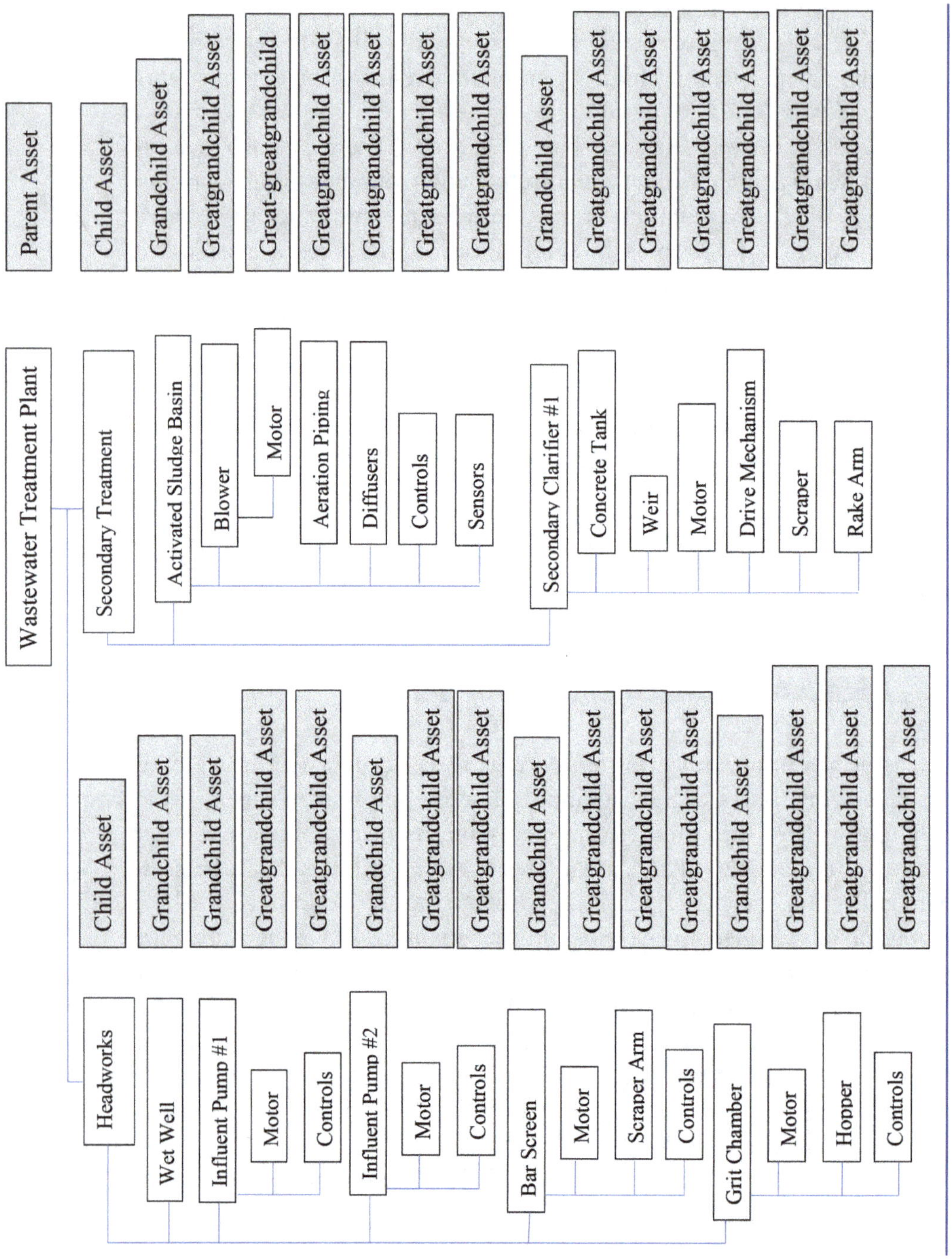

FIGURE 3.3 Parent–Child Relationships

- Location
- Condition
- Useful Life Remaining
- Replacement Cost

Additional data that can be collected may include the following:

Size
Installation Date
Manufacturer
Manufacturer's Contact Information
Model Number
Serial Number
Supplier Name
Supplier Contact
Operating within design standards
Initial Construction Cost
Under Warranty
Warranty Expiration
Operational Status
Material
Redundancy
Energy using asset (Y/N)
Type of energy use
Connected to emergency generator

A specific asset class will have other, more asset-specific, attribute data that can be collected. As an example, consider a fire hydrant. Additional attributes can include:

- Wet barrel/dry barrel
- Presence of an isolation valve
- Color of body
- Color of top
- Designated for use by commercial entities wanting to get water
- Last date of flow testing
- Flow test results
- Expected flow rate
- Expected pressure
- Last date hydrant was used to fight a fire
- Pipeline connected to
- Size of line connected to
- Street address

Some of the asset attribution information can be collected from a field inspection, while other data will need to be collected from operating the asset or from outside sources, such as design drawings. The attribute data collection can be phased such that only the field data is collected first. Then when the asset is actually operated within the routine course of business or during testing or maintenance activities, the additional attribute data can be gathered. For example, it is possible to go to each fire hydrant in the system and gather the data that can be obtained from laying eyes on the hydrant. Then when the annual fire hydrant testing is completed, additional data can be collected to further the inventory. The additional data that can be obtained from outside sources (such as a fire department who tests hydrants on a routine basis or who keep data on fire hydrant usage during fires) can be added at any time.

Sources of data for assets, other than from field investigations or operating the asset, can include:

- Aerial photographs
- As-built record drawings
- Existing utility maps
- Repair, maintenance records
- operation and maintenance (O&M) manuals
- Record drawings
- Inspection records
- Facility plans
- Condition assessment reports
- Engineering reports
- Master plans
- Photographs
- Manufacturer's information
- Discussions with past staff
- Discussions w/engineers or contractors that previously worked on the system
- Fixed asset reports

Before collecting data on assets, it is a good idea to form a group of stakeholders within the utility to brainstorm what asset data would be valuable at all levels. For example, the internal stakeholders can include: operators, maintenance personnel, supervisors, managers, engineers, finance officers, customer service personnel, meter readers, or lab technicians. While the basic information needed may be the same for all, it is also likely that each of these groups may have additional data requirements to meet their needs. Forming a group at the beginning of the process can help ensure that the correct data is gathered at the beginning, rather than having to go back out several times to collect more data.

It is worthwhile to examine what type of information is gathered for each class of asset periodically to make sure you are neither missing important attributes nor collecting unimportant ones. This type of review could be done on the same schedule as the revision of the asset management plan (every 5 to 10 years) or another cycle if advantageous to the organization.

In organizations that use geographic information system (GIS) software that are built on a standard data model (e.g., Esri), the programs may already contain a suite of fields (attributes) for asset data that has been found to be useful over a wide sample of utilities. It is worthwhile for the stakeholder group to review the models and ensure that they include the attributes that are appropriate to their facility and make any necessary adjustments. It is important not to assume that the already populated fields are the correct ones for each entity.

4.0 COMPLETENESS OF THE ASSET INVENTORY

No matter how much effort the utility puts into its asset inventory, it is unlikely that it will ever be complete or entirely accurate. It is necessary to get comfortable with the idea that there will be gaps in the data and possibly errors. Over time, many of these gaps can be addressed and errors corrected, but some will persist for a long period of time and may not be eliminated until the asset is removed from service and a new asset is installed.

The fact that the inventory will not be 100% complete or accurate does not mean that effort should not be put into gathering the data. It is most important to have a plan in place for adding additional data over time and to consider the cost/benefit analysis of collecting various types of data. Some information is more likely to be worth an immediate investment of effort to collect while other attribute information is less likely to have that effect. The plan to collect the information should include business processes for data acquisition, maintenance, editing, and publishing for both new and existing assets, as these procedures may be different.

Below are some examples:

- *Visible plant assets:* Assets that are visible and within treatment plants, pump stations, lift stations, and booster stations are relatively easy for data collection and the information is valuable and useful. These assets should be put into the inventory and as much data as possible should be collected on them.
- *Visible field assets:* Assets that can be seen in the field, such as hydrants, manholes, meters, storage tanks, pressure reducing valves, and air relief valves can be relatively easy for data collection, but there are generally a lot of these types of assets in the system. Therefore, data collection can be quite time consuming. A large system may have thousands or tens of thousands of these types of assets. It may be too resource intensive to collect all the data on these assets in the first year of an asset management program. Furthermore, the benefit of having data that quickly may not outweigh the cost associated with the effort. Instead, a plan to gather the data over time may be more effective. During the course of a year, operators and other personnel will have occasion to visit these assets during routine operation, inspections, repair events, testing, or other activities. While at

the asset, information can be gathered to populate the inventory. For example, some utilities ask meter readers to gather data on a percentage of the meters each time they go out to do meter reading until data is obtained on all meters. This process may take a number of years, but over time, the inventory will be more complete.

- *Buried field assets—gravity wastewater pipes:* There is a bit of difference between water and wastewater pipe, and gravity and force mains so these will be discussed separately. Sewer manholes in gravity pipes can provide an opportunity to gather some information about the collection pipes. However, this method will not provide condition information. It takes time to remove the lids of manholes so in a large collection system, considerable time and resources would be required to gather this information. Therefore, this is a part of the inventory that should be built over time. As pipes are investigated with closed circuit televising equipment (CCTV), information should be added to the inventory. A schedule can be created to proactively check a certain length or number of pipes per year to gather basic inventory and condition information so that the inventory can be completed over time. For example, if the system had 100 miles of pipe, they could proactively check an additional 5 miles of pipe per year that was not otherwise slated for CCTV work, such that a complete inventory (or relatively complete) could be created over 20 years. This rate of adding information to the inventory could be adjusted higher or lower as the benefits of the information outweigh the costs of obtaining it or the costs outweigh the benefits.
- *Buried field assets—wastewater force mains:* Force mains are buried and more difficult to locate because they are not included in the utility's CCTV work. If there are prior records, such as as-built drawings, the location can be estimated off the drawings. It is not recommended that the locations be determined using pot holing unless there is some other reason the location must be determined, other than for asset inventory purposes. If there are no records, the locations will need to be estimated from past experience or knowledge of where the lines are expected to be or through repair work.
- *Buried field assets—water distribution pipes:* If there are as-built drawings or similar records, water pipe information can be estimated based on this information. It can take some effort to translate the information into the asset inventory so it may need to be done over a period of time. If these records do not exist, water pipes can be difficult to locate because they are buried and the surface clues—valves, meters, hydrants—help identify a pipe is nearby but do not

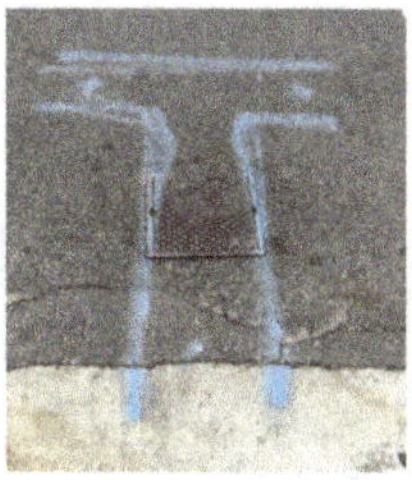

necessarily identify the exact location of the pipe. For example, a valve shows a pipe runs through it, but not necessarily the direction the pipe runs. A fire hydrant shows that a water line is near and possibly the general direction based on seeing multiple fire hydrants but does not identify exactly where on the street the pipe is. Beyond the location, there may be little that can be gained from the street level unless it has been marked in the past. Some examples of surface indications (some are temporary while others may be permanent) of pipe locations, sizes and types are shown in the pictures. A great time to add pipe data is when pipes are being repaired. Another good time to add data is when the pipe is being constructed, replaced, or commissioned. At these times, the pipe's size, location, type, and condition can be observed. Waiting for this information can take a while to complete the water line inventory so patience will be required.

- *Buried field assets—water isolation valves:* Many valves are visible from the street because there is some type of valve lid. However, some valves get paved over, making it hard to access and find the valves. The location of valves may be relatively easy to obtain (when they are visible) but there may be hundreds, thousands, or tens of thousands of valves in a system depending on how big the system is. Therefore, it can take a bit of time to locate them all. Collecting additional information on valves, such as number of turns, direction of turns, operational status, normally closed or open status, takes additional effort, which may need to be done over time. Data can be collected when valves are operated as part of repair activities.

A cost benefit analysis can be performed to balance the benefit of having each individual type of information with the resources (time and money) needed to obtain it. For visible assets or assets for which there are few in the system, it is generally worth the effort to gather the information quickly. When there are many, many assets or the assets are not visible, the calculation may shift to the point where the cost of gathering the data all at once may outweigh the benefits.

5.0 CONDITION ASSESSMENT

A condition rating is a necessary attribute for all assets in the inventory. The condition defines the physical state of the asset at this moment in time. It can help inform the useful life remaining (discussed in the next section); operational status; needed maintenance interventions, including repairs; and necessary replacements. Reviewing the collective condition ratings for all the assets can also provide a bigger picture assessment of the facility as a whole.

A simple condition rating system usually uses something like a five-point scale that can either be numeric or word-based. The disadvantage of a numeric rating is that it can be confused with the numeric criticality scores that will be associated with the assets (discussed in the next chapter). An example of a five-point scale is shown in the table below (Table 3.1).

To develop the rating system, it is necessary to establish the conditions that would be present for each class of asset for each rating. The easiest way to start this process is to consider the boundary conditions—what would cause the rating to be a 1/excellent or a 5/poor? Then fill in conditions for 2/good to 4/fair in between the two boundary conditions. As an example for isolation valves, see Table 3.2.

It is possible that not every condition exists in the numeric rating, so an operator needs to decide which conditions are most important. For example, the valve may not operate, but the valve can is not full of debris. The fact that the valve does not operate makes the asset a 5, even though the valve can is not full of debris. The example in Table 3.2 is meant to demonstrate the process and should not be considered the definitive condition rating system for valves. This rating should be modified to fit the utility's particular needs. Additional criteria can be added to describe the conditions.

Using this rating scale, an operator can evaluate each isolation valve and apply condition ratings. The ratings scale should be as objective as possible; it should be reasonably easy for different individuals to assess a valve but come up with similar ratings. It should also be easy to rate valves across the entire utility and give them ratings on the five-point scale. Unless the

TABLE 3.1 Condition Ranking Criteria

Numeric Rating	Word Rating
1	Excellent
2	Good
3	Average
4	Fair
5	Poor

TABLE 3.2 Condition Ranking Criteria for an Isolation Valve

Rating	Description
1/Excellent	Valve opens and closes with normal effort. When valve is closed, water stops flowing. When valve is open, water flows as intended. No visible signs of wear on any visible components. Operators have not identified any problems with the valve. No defects noted. The valve can was clean.
2/Good	Valve opens and closes with slightly more effort than normal. When valve is closed, water stops flowing. When valve is open, water flows as intended. No visible signs of wear on any visible components. Operators have not identified any problems with the valve. No defects noted. The valve can may have been full of debris and had to be cleaned out or it may be clean.
3/Average	Valve opens and closes but needs additional effort. When valve is closed, water stops flowing. When valve is open, water flows as intended. Some visible signs of wear on visible components, but not excessive. Operators have not identified any significant problems with the valve, but may have minor issues with it. No defects noted. The valve can was full of debris and had to be cleaned out.
4/Fair	Valve opens and closes only with excessive effort. When valve is closed, water still flows through. When valve is open, water flow may or may not be as intended. Visible signs of wear on visible components. Operators have identified problems with the valve in the past. Defects noted. The valve can was full of debris and had to be cleaned out.
5/Poor	Valve does not operate (if already open, it does not close; if already closed, it does not open). Operators have identified problems with the valve in the past. Defects noted. The valve can was full of debris and had to be cleaned out.

system is relatively new, it is highly likely that at least some valves will be in each of these conditions.

It is important to avoid the temptation to score every asset a 3 or average. It might feel that this rating cannot really be considered wrong and may not be questioned, whereas ratings at either end of the scale might be. However, if every (or most) asset(s) is a rated 3/average, there is no point to a condition rating system as it will not inform where or when interventions should occur or how well past interventions have worked to maintain condition. It will not separate assets into categories of how much effort they might require. If all assets are rated the same, they would all deserve the same amount of attention, which is very unlikely to be the case.

Likewise, the temptation to rank every asset a 4/fair or 5/poor for a facility that, as a whole, is considered to be in bad condition should be avoided. Even in cases in which a facility may appear to be in a bad state of repair, there are going to be many assets that are not in fair or poor condition. There can be a tendency to assume that if the plant has violations or has been poorly managed in the past that everything is bad. It is still important to go through the plant with an objective eye and rank condition. This type of ranking will help the facility determine where to intervene with maintenance or repairs and how to salvage as much of the facility as possible.

Certainly, more effort can be put into the condition rating process than described here, but it is worth asking the question, is the more refined condition rating worth the effort it will take to obtain it? Will it help manage the assets in a more cost-effective, efficient manner? Will it provide data that will help inform decisions that would not be available with a simpler assessment? It will take time and resources to perform a more complex condition rating that is based on more factors and more involved data collection, so the benefit of the data must outweigh this cost. There is an opportunity cost associated with the more advanced practice. If, however, there is value to a more complicated system, this type of rating structure should be developed and applied.

The condition ratings should be as objective as possible such that two different people rating the asset would be likely to get the same or similar rating. While not everyone might give the exact same score, it should not be possible for one rater to give the asset a 4 and another to give it a 2. Some assets can have condition ratings primarily from operator knowledge and field observations, while other assets will need to be supplemented with additional data or operation of the asset.

Condition rankings can be used for several purposes including those on the list below.

1. Identify assets that are underperforming.
2. Predict when the assets are going to fail to deliver their intended level of service.
3. Help determine the reasons for performance deficiencies.
4. Determine the type and timing of maintenance interventions or corrective actions.
5. Identify failed or nearly failed assets to help with capital improvement planning and data-informed decision-making.

If condition ratings can provide worthwhile information for the utility, how much effort is necessary to obtain the ratings? For the most part, simple condition ratings will be sufficient. Between the knowledge of personnel familiar with the assets, field observations, and any additional information obtained by operating the assets, the simple ratings will most likely be sufficient.

In some cases, additional effort may be worthwhile, such as highly critical assets. (Criticality will be discussed in Chapter 4.) For this small subset of assets, additional techniques can be used, such as a diver in a storage tank or a closed-circuit TV camera in a sewer line to gain additional information to refine the condition rating. Other technologies can be used, such as vibration analysis or thermal analysis to gain insight into asset condition.

Another option is to choose a random subset of assets and perform a more aggressive condition assessment on those assets and then apply the results to similar assets in the system. This approach can work well with assets for which there are a large number in the system but the actual condition rating is difficult (e.g., customer meters and pipe). In the case of customer meters, if there are thousands of meters in the system, it would be very time consuming and difficult to give each meter a condition rating. However, the meter inventory could be examined to see what categories the meters fall into, based on size, type, manufacturer, date of installation, average flow-through, or other characteristics. Then a random, statistical sampling of the meters could be conducted to evaluate the condition. This condition rating could then be applied to all the meters in that particular category until such time as more detailed information on the actual meter condition could be determined through normal operation and maintenance. For example, a meter reader might note a leak or a stopped meter or a customer complaint might trigger an investigation into the meter condition.

Given that the condition ratings describe the physical condition of the asset at this moment in time, the ratings will not stay the same for a long period of time. As the assets deteriorate over time or experience conditions that can cause failure, the condition rating may decrease. If maintenance interventions are performed to improve the condition, then the rating may increase. It is necessary to reevaluate condition ratings for the system as a whole on a periodic basis and on specific assets after any interventions that change the condition. It is best if condition ratings can be reevaluated annually, but if that is not possible, then at least every 3 to 5 years. If the condition ratings are revised during routine, corrective, or preventive maintenance activities, it is not necessary to redo the ratings for those assets as part of the overall update. This approach can greatly reduce the need for a routine reevaluation of condition and is a good overall management practice. In this type of operation, the condition rating is reviewed, and revised as necessary, as part of the work order.

It may be useful to consider the failure modes of assets when thinking through condition assessment. A further discussion of failure modes is included in Chapter 4, but a brief description is provided below.

- *Mortality:* The physical failure of the asset due to deterioration or destruction that prevents the asset from performing its intended function. The asset may be repairable, but it will no longer function properly until it is repaired or replaced.
- *Level of Service:* The asset can continue to perform its function but not at the level of service required by customers.
- *Capacity:* The asset does not have the physical capacity required to meet the demand or the needed capacity exceeds the design capacity for the asset.
- *Financial Inefficiency:* When it costs more to continue to operate and maintain an asset than it would to replace the asset or the cost of continued operation and maintenance is a large part of the overall asset cost.

An asset can fail for any of these reasons individually or it can fail for multiple reasons. The consideration of failure modes can also assist with the consideration of useful life remaining.

While it is best if the utility personnel perform the condition rating process, because they know the assets best and it is important for any learning about the assets that takes place during this process to reside with the facility personnel themselves, a utility may choose to hire a consultant to perform this work. If this is the case, there are a few considerations to keep in mind.

- Ensure that the consultant uses a process created in conjunction with the utility personnel, especially operations and maintenance personnel. The final condition rating system used by the contractor must be approved before it is used.
- Test the rating system between utility and non-utility personnel to ensure that the ratings are consistent. If the two sets of personnel do not agree, discuss the differences and determine a methodology to reach a similar rating.
- Involve utility personnel as much as possible in the process.
- Ensure the contractor shares any concerning findings about particular assets immediately upon discovery (i.e., an asset that is vibrating badly or leaking should be identified).

- Specify, before any work starts, the format of the data both in written and electronic format. Ensure that the data is in a compatible format so that it can be added to the utility's software system (if using one).
- Have the contractor provide a briefing for utility personnel to discuss the findings.

6.0 USEFUL LIFE REMAINING

Useful life remaining is a concept that indicates how much longer the asset is able to do the job for which it is intended; it is the amount of time until the asset reach its minimum acceptable service level. The useful life remaining is not the same as an engineering estimate of how long an asset will last based solely on age. Age, by itself, is not a very reliable determinant of how long something will last because many other factors affect whether the asset reaches the end of its life sooner, later, or on this time scale. Factors can include: conditions the asset is exposed to (e.g., gases, sea air in coastal communities, soil conditions), materials of construction, maintenance history, failure history, weather conditions, or quality of water inside or outside of pipe. Using age-based estimates alone is highly likely to underestimate the length of time the asset can perform its functions, which will in turn affect the ability to create an accurate long-term capital improvement plan and an associated replacement budget.

The concept of useful life remaining is shown in Figure 3.4. The y-axis shows condition while the x-axis shows time. Over time, the condition of the asset will deteriorate along some kind of curve (the solid blue line). The curve is unlikely to be a smooth deterioration such as the line shown here, but the idea is the same no matter how the asset deteriorates. At some point, the asset's condition will deteriorate to the point where it is no longer able to meet the minimum acceptable service level (the dashed blue line). This level is different for each asset and even can be different for different systems. Some systems may have higher or lower tolerances for minimum acceptable service level. Some examples of assets and potential minimum service levels are:

- Fire Hydrant—Cannot meet fire flow requirements
- Isolation Valve—Cannot hold flow back when closed or impedes flow when open
- Water Main—More than three breaks per segment over 5 years
- Well—Quantity of water produced is 20% or more below design flow

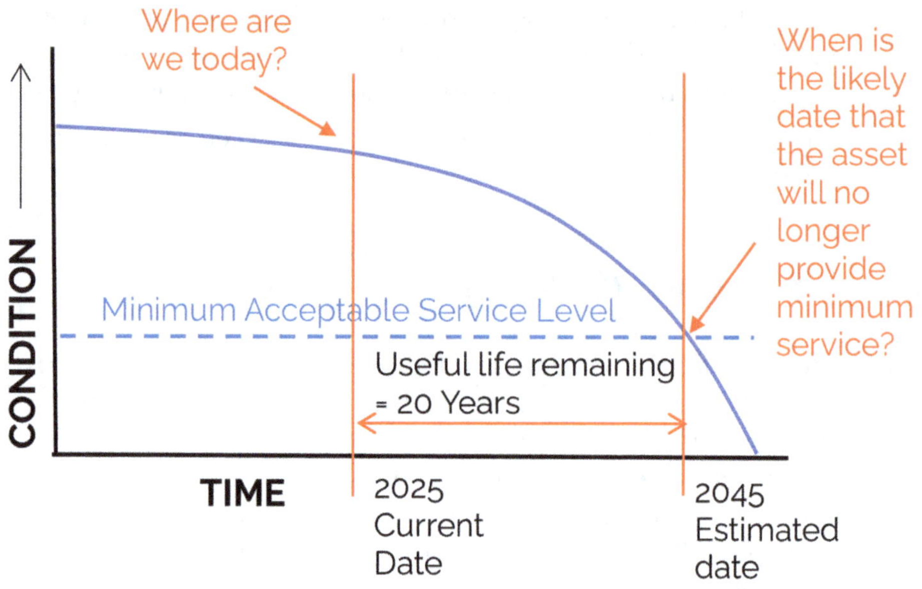

FIGURE 3.4 The Concept of Useful Life Remaining

- Flow Meter—Accuracy of less than 97% either above or below expected flow

These levels are examples, and a system can decide what levels should be set as the level below which the asset is no longer able to remain in service without some type of intervention.

Once those minimums are determined, the next step is to estimate when this is likely to happen. This is the system's best estimate based on all it knows about the asset. Taking into account the maintenance history, the failure history, the number of repairs, experience with similar assets, the amount of service (i.e., how much flow or how much time in service), the environment the asset is subject to (assets in harsh environments are likely to fail sooner), and any other pertinent factors, the system personnel should make their estimate of when they expect the asset to no longer be able to do its job. This is the vertical red line to the farthest left side of the graph. In this case, the year is estimated to be 2045. The next step is to determine where the asset is in time today. This is the vertical line to the left side of the graph. In this example, 2025. The useful life remaining is the time between the two vertical lines, in this case, 20 years.

Because the date at which the asset will reach the minimum acceptable service level is an estimate, it is logical to ask the question, what if the guess is wrong? The answer is that these estimates are not the definitive answer regarding when an asset is going to be replaced. Assets will be reviewed in more detail when they are placed on the short-term (1–5 years) capital improvement plan (CIP). Furthermore, the estimates can always be revised as new information is learned about the assets. Because these estimates will be used to place items on the long-term CIP (6 to 50 years or 6 to 100 years), the estimates are not expected to be as reliable for the later years (past 20 years) as estimates of useful life for assets in the short term. As projects come up for funding and move to a conceptual or detailed design, more consideration should be given to the expected service life.

For assets in which the useful life remaining is estimated to be less than 5 years, there should be a reasonable amount of information upon which to base the replacement decision. For example, failures have occurred, parts are hard to obtain, deterioration in the material the asset is made of has occurred, or a physical attribute shows problems (e.g., vibration, noise, leakage). For assets that are expected to last 20, 50, or 75 years or more, it is expected that adjustments will be made to the useful life over time as more information is learned. The estimates may be either direction—lengthening the life or shortening it—depending on how the assets are performing.

Another consideration is the fact that basing the useful life on an engineering estimate of how long a particular type of asset is expected to last based on age is no more accurate and very well could be less accurate than the system's best estimate of the useful life. These types of estimates would require the same type of adjustment over time, but there may be more reluctance to do so because they appear to be based on something official. These types of estimates are particularly problematic for long-lived assets. Because every installation has so many factors that make it unique (soil conditions, atmospheric conditions, installation techniques) and the time to see the effects of failures is so long, it is difficult to make better estimates of this type than a utility's best judgments.

A medium-sized wastewater utility began an asset management process and took an inventory of all of the wastewater assets in their treatment plant. The plant was running well and meeting all permit limits. It was not scheduled for any major renovations, replacements, or upgrades. The personnel used standard engineering estimates of how long different types of assets were expected to last and the results indicated that 75% of the assets were past their useful life. This result concerned the city manager as there was no plan to make that type of investment (to replace all or most of the wastewater plant). The problem was that the results did not match the reality of the plant. The assets had been well cared for and routine and preventative maintenance were performed on schedule. When the personnel went back and redid the estimates using their best judgments, the percentage of assets beyond useful life dropped to 10%. These assets were primarily in this category due to obsolescence. Although the assets could perform, it was no longer possible to get parts. This result was within the city's ability to address.

It is not always necessary to replace an asset that has reached the end of its useful life. The underlying asset may be able to be refurbished instead. Some examples of this type of approach would be lining a pipe, lining a storage tank, cleaning a well, or working an isolation valve to get it moving again. As an example, a pipe asset may be scheduled for relining in 5 years with a useful life remaining of perhaps 10 years. When the pipe is relined, it may provide another 50 years of life in the pipe so the useful life remaining can be updated accordingly. Some assets can be refurbished multiple times, while other assets can only be refurbished once. The amount of additional life will be less for subsequent refurbishments than the first time because the underlying asset is continuing to degrade.

Useful life remaining is not the same as total expected useful life. Useful life remaining refers to the amount of time left in the asset's life, not the total time the asset has been or will be doing its job. The total expected useful life can take into account the entire life of the asset from cradle to grave and can include refurbishments that are expected to be completed along the way.

Useful life remaining estimates are not a once-and-done activity. The useful life remaining should increment each year and should be reevaluated periodically when assets are undergoing inspections or major repairs. It is important to keep this information current.

7.0 ASSET MAPPING: IT'S A PROCESS

Locating assets is a very important part of the overall asset inventory. This location is especially important for what are called horizontal or field assets. These assets are in the community, rather than a treatment plant, and include many assets that are largely unseen. Examples of horizontal assets include: pipe, isolation valves, service lines, hydrants, pressure relief valves, air relief valves, pumping stations, manholes, service lines, clean-outs, meters, and storage tanks/reservoirs.

Locating horizontal assets is particularly important for the following reasons:

- *Emergencies:* It is critical to find an asset quickly when emergency conditions occur, such as fires.
- *Legacies:* Long-term operators often have extensive knowledge of asset locations in their heads but not written down or in an electronic file. Putting this information in a format that can be shared creates a legacy from more experienced operators to newer, less-experienced ones (at least less experienced with that particular system).

- *Repairs:* If repairs need to be made on underground assets, it is more efficient to have good locational information of that asset and any additional assets that are needed for the repair (such as isolation valves, manholes, service lines, or meters). Operators can waste a lot of time locating assets when the time would be better spent working on the repair.
- *Capital Improvement Planning:* It is important that capital improvement planning includes horizontal assets. These assets may be forgotten as part of the process because they are out of sight and more difficult to assess.
- *Trending of Operational Data:* Some operational data is best understood in a spatial way. For example, if pipe breaks are plotted on a map, trends may be observed that indicate parts of the system more susceptible to breaks due to soil types, groundwater, material, size, depth of bury, or any other distinguishing characteristics. The map can also be used to plot sample results that might indicate a problem with operations, dead ends, raw water quality, or other issues.

The mapping process involves five steps as shown on the diagram below (Figure 3.5). The steps are described in more detail below.

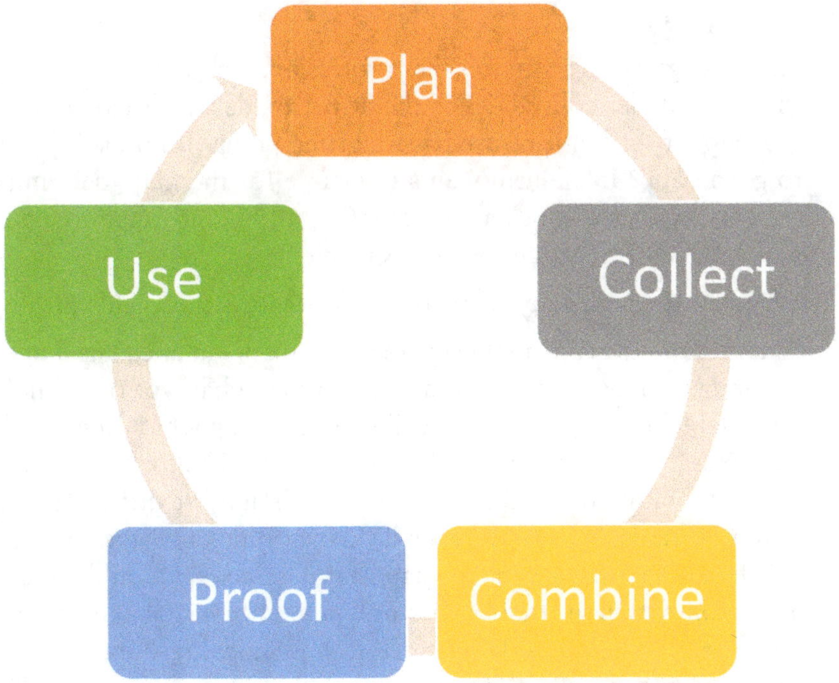

FIGURE 3.5 Mapping Process

Plan: Decide what information to collect, what tools and/or software will be used to collect the information, the time frame in which to collect it, and who will do the collecting. The planning process should include all pertinent staff/stakeholders—or their representatives—who would be involved in the mapping process to ensure all viewpoints are considered. Some examples of items that should be considered during the planning process include:

- Will data collection be done for both horizontal and vertical assets or just horizontal? Will the process be different for horizontal and vertical? In what ways? Will it be different for different classes of assets? Will water and wastewater assets be collected at the same time?
- What classes of assets should be included on the map?
- What type of map will be created? What will it be used for? Who will need to use it? Will it be used electronically or in paper form or both? Will it be used on devices out in the field (e.g., tablets or phones)?
- What information already exists to help create the map? Is there already information in a map and are you just adding to it? Does another department have a base map of the utility that can be added to? Are there as-built drawings that can be used as a starting point?
- If there is a current map available, has it been field verified? If not, is that the main purpose of mapping activities (i.e., to field verify an existing map)?
- Has an engineering firm or other professional entity create a map for the utility under contract? Did the firm provide an electronic copy of the map? If so, can it be used as a starting point? If not, is it possible to get a copy? If the firm wants to charge for the map, determine if the cost savings of not having to gather all the information again is worth the cost to purchase the electronic data.
- What are the highest priority assets to either map or field verify?
- What resources are available to create the map—financial, technological, and personnel? It is important to consider what type of technology is available or could be made available, including what licenses might be necessary and how much these might cost. The mapping exercise should fit within the available resources or it may be difficult to finish or it may not be possible to keep it up over time once it is completed. It is better to have a simpler map that is relatively complete than a more complex map that covers only a small part of the system or just a few asset classes. In terms of technological resources, the type of equipment to collect locational data should be considered. There are simple, inexpensive, less accurate ways to collect the latitude and

longitude, such as a smart cell phone, which is essentially free with the service you are already paying for, mid-range technologies that are moderately expensive that give reasonably accurate results, and at the other end of the spectrum, there are very expensive technologies that are extremely accurate.

- The need for accuracy may vary between asset types. Those assets that can easily be seen on the surface may need less accuracy than other assets, such as customer meters in mild climates that are located near the property boundary and have visible meter lids. Assets that are harder to locate and are very important to proper operation may require more accuracy.

A few brainstorming sessions would be a good way to get the planning started followed by the creation of a written mapping plan that spells out what will be done, how it will be done, who will do it and the schedule by which it will be completed.

Collect: Once the planning is completed, it is time to collect the information. During this process, the locational information (lat/long and potentially street address as well) is the most important part of this data collection because it will be used to produce a map. Additionally, all pertinent information about the assets should be collected at the same time. The technology(-ies) chosen in the planning process should be used for the collection of data. It would be worthwhile to beta test the approach with a few assets to make sure the process will work as intended before proceeding to invest considerable time and effort. There may be ways to increase the efficiency of data collection that were not initially considered.

> One utility initially intended to take laptops into the field to do data collection on visible assets. However, that process proved to be slow and cumbersome. They learned that it was easier to snap a few strategic photos of the assets that visually contained all of the information they needed and then enter the data in the office using the photos.

There are many options available now for data collection. It may also be necessary to pair up two technologies or two options—one for attribute collection and one for locational data. For example, a global positioning system (GPS) device may be used to collect the location while a phone or tablet may be used to collect other attribute information related to the assets. The phone or tablet can be pre-loaded with a data collection app that has pre-filled drop-down menus and contains all the desired attribute fields.

There are various options for collecting GPS data that range in accuracy and cost. At the low end of accuracy and cost is the use of cell phone or tablet technology to get a GPS reading. Handheld GPS devices will be a bit more expensive and prices will vary based on the accuracy of the device and the features included. At the high end of the price spectrum will be survey-grade GPS that will be very accurate but can also be costly. It is necessary to balance cost with the need for accuracy and choose the approach that best serves the utility's needs. It is also possible to have some assets located more accurately than others if the need for information varies.

There are side benefits to mapping activities and data collection. There are many things that can be learned when you set eyes on the whole system. For example, you may find an asset that is not in proper working order that you were unaware of or anomalies in the system (e.g., a place where there should be a valve but none are visible) or missing assets (e.g., buildings that should have a meter but do not).

Combine Sources of Data: Maps are more than a way to tell where something is, they can also tell a story. Combining sources of data into the map can help create an image or story. It can also help move the utility from a place where someone knows a bit of information themselves that is not widely known outside that person to a place where knowledge is available to all. See Figure 3.6 below.

Make Maps, Use and Improve Them: The most important thing with maps is to actually use them. A map is most beneficial if it is an accessible format that all can use. That format might be hard copy for some employees and electronic for others. In some cases, the maps may be accessible in real time in the field while in others connectivity may only allow maps to be used in offline mode.

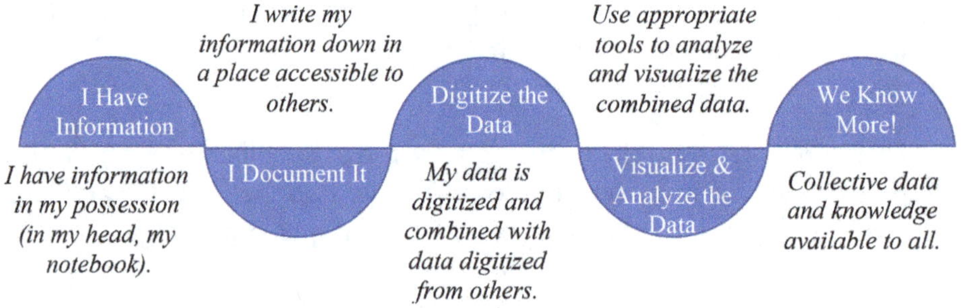

FIGURE 3.6 Combining Sources of Data

It is very likely that the utility map is not entirely accurate. It is common to have mistakes and issues with location or attribute data, especially with assets that cannot be seen. Initially, a best educated guess regarding the type, size, or location of pipe is good enough. It is generally not worth the effort to pot-hole distribution pipes to find more accurate information, especially for small diameter pipes in residential areas. In some cases, the pipe may be so critical that it is worth this effort or very exact locational information may be extremely important.

However, over time, repairs may require the distribution pipe to be dug up. At this point, utility personnel will lay eyes on the pipe and be able to determine with certainty the pipe material, its location, and its diameter. They will also be able to provide a condition rating. At this point, the map can be updated with field-verified data and the accuracy will increase.

Another opportunity to locate pipes is when the utility is requested to locate drinking water, sewer, or stormwater pipes as part of the state's one-call system. If a utility employee is required to find the location of a line that is not currently on the map, the pipe should be added once it is located. Once a reasonably accurate map is created, participating in the one-call line location process becomes easier.

The requirements under the Lead and Copper Rule related to service line inventories required drinking water utilities to determine if any of the service lines were lead. In cases in which the lines could not be definitively identified as either lead or non-lead on the initial inventory, pipes had to be marked on the inventory as "unknown." Following this categorization, utilities had to follow protocols to determine the pipe material of the unknowns and move them out of that category. One methodology involved digging up the pipes to make a definitive assessment. This information, along with all other service line inventory data, needs to be included in the asset inventory. While the potholing can help identify service lines, it is generally not sufficient to determine the distribution pipe type (i.e., the potholing is generally on private property in this case, not in the street where the distribution piping is located).

Vertical Asset Mapping: Mapping is generally completed for horizontal, or field, assets. However, there are types of maps that can be created for vertical, or plant, assets. A map can be created around buildings or portions of buildings and the vertical assets can be mapped according to their building locations. The asset locations can be considered "field guides" that can aid the operators in locating the vertical assets when they receive work orders. The work orders could include a location that is keyed to the vertical asset map. A simple example is shown on Figure 3.7 below.

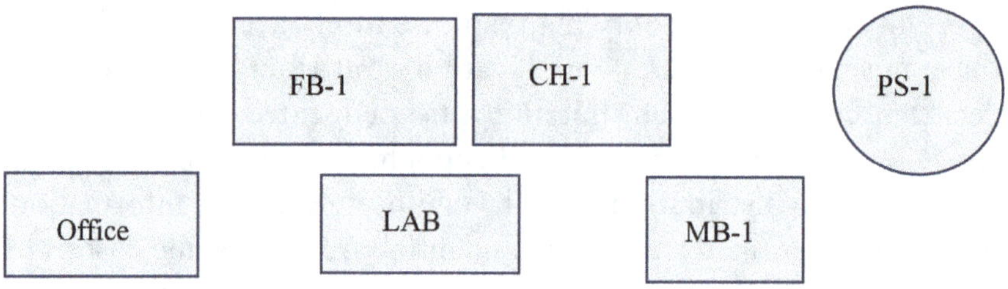

Key:
MB-1: Maintenance Building
FB-1: Filter Building
CH-1: Chlorination Building
Office: Office Building
Lab: Lab Building
PS-1: Pump Station

FIGURE 3.7 Vertical Assets Location Map

A field guide lists the major assets or groups of assets in each location. It can be as detailed or not as the facility desires. The guides can be created in handout form, as a laminated sheet, or could be used on an electronic device (phone or tablet).

Internal vs. External Map Creation: A map can be completed by the utility personnel themselves or by individuals associated with another entity of which the water utility is a part (i.e., a municipality or a county). Seeking assistance from others within the larger community can be a very efficient way to complete a map. These entities may be more knowledgeable in mapping technology, may already have base map information that covers the utility service area, and may already pay for appropriate mapping licenses. One option is to use water system personnel to gather the asset locational data and then provide that data to the personnel who make and revise the maps. These individuals can keep the map current over time as well and help the utility personnel use the map.

A utility in the northwest U.S. worked with the planning department to create a map of the utility. The planning department had the base map that included streets and parcel data. The GPS locations of all the water and wastewater assets were collected and provided to the planning department. The planning department was then able to serve the mapping data to the water system personnel. Having a more complete map helped the planning department and having an entity to make and keep the map helped the water utility.

If the map is created by the utility, someone within the utility should be designated as the person in charge of the map. That person needs to have the skills and knowledge to run the mapping software, the ability to help others use the map, and the capability of updating the map over time.

Another option is to contract with a local high school, community college, or university to help with the mapping. These individuals may be able to help collect the data or they can make a map from data collected by the water utility personnel. The water utility can come back to the school/college/university on some type of regular basis to have mapping updates completed.

A utility in the Rocky Mountain region of the U.S. worked with a local high school that included a class on GIS mapping. The utility collected GPS locational data and then the data was provided to the high school class who created the map. The class can update the map on an annual basis for the utility. In the period in between, the map will largely be used in paper format.

The Environmental Protection Agency (EPA) has been providing grants on an annual basis to assist small drinking water and wastewater systems (small drinking water systems in this context are those serving less than 10,000 population and small wastewater systems are those treating less than one million gallons per day [MGD]). These training and technical assistance grants are currently awarded to multiple entities, any of whom can provide different levels of assistance to small water and wastewater systems (as defined above) with creating a map. These resources should be considered if you need help with mapping or getting started with the process.

In some states, there is additional assistance, beyond the EPA funded entities, to create maps or there may be grants that fund this activity under the Asset Management grants that several states provide. One example is the state of Vermont that provides planning grant assistance for asset management that can include making a map.

If the entire mapping process needs to be contracted out to an entity outside of the utility, it is recommended that the utility personnel stay involved in the process. There is much to be learned from mapping activities and it is best if the utility personnel are part of the learning process. Additionally, it can be hard for outside personnel to locate buried assets. The most important thing to remember if the map is being completed by a contractor is to ensure that the utility owns the data. If this is not explicitly stated in the contract documents, it is possible that the utility will get a printed map or access to an online map but not access to the full set of data. Even if no one from the utility is well versed in mapping or mapping software, the utility still needs to maintain ownership of the data. If not, it is possible that if the utility wants to use the map at a later date or add or revise data, it may have to pay the contractor for that use or pay to have the changes made. The fact that the utility intends to own the data should also be clearly stated in the request for proposals (RFP) so that anyone bidding on the project knows this is the ultimate disposition of the data and can price their response accordingly. The utility should also request the data be exported in a format that would be compatible with most mapping software.

Several utilities over the years have had a map created by an outside entity in the past but the data resides outside of the utility with that contractor. In some cases, the contractor agreed to provide the data to the utility so that it could be part of the asset management inventory process but in other cases, the data was not able to be obtained. In one case, the data no longer existed at the contractor's office. In another case, the utility knew someone had created a map but could not remember who did the work. Even though they contacted all the engineering firms they had worked with, the data could not be found. Only a paper copy was available.

No matter which process a utility uses for mapping (internal or external), it is necessary for the utility to develop a business process around how mapping errors will be documented and communicated, how maps will be updated, and how and when maps will be published and communicated to all staff. This process should explain which individuals within (or outside) the utility are responsible for the various portions of keeping the assets updated.

8.0 ASSET INVENTORY SOFTWARE: A TOOL, NOT THE SOLUTION

Asset management is not a software program; it is important to remember that asset management is a thought process in which data is used to help systems make better decisions. However, strategically employing various technologies can assist a utility in handling data sets and facilitate data analysis and data-driven decisions. There are several technologies that might be beneficial, such as supervisory control and data acquisition (SCADA), geographic information systems (GIS), financial databases, automatic meter reading, but the focus of here is technologies that can be used to store asset inventory and maintenance data. These software programs are referred to as a computerized maintenance management system (CMMS) or asset management software (AM software).

While extremely small systems can make use of inexpensive programs, such as Microsoft Excel^{TM}, larger systems may wish to have a more sophisticated software program developed by a third-party vendor. This section helps utilities determine if they need more sophisticated software and discusses the various aspects of the software selection process.

Many utilities, especially small utilities, may be unsure whether CMMS/AM software would benefit their operation. Below are some guiding questions that can help a utility decide whether to take the next step in their CMMS/AM research or determine that operations can continue without one.

Record Keeping: What does record-keeping look like? Is it chaotic, with related material kept in many different locations? Is it hard to locate assets? When record-keeping is disorganized, utility personnel may spend more time finding relevant information than making repairs. CMMS/AM software is capable of storing everything in a single, searchable database that facilitates informed data-driven decision making. It also improves the tracking and scheduling of labor, helping identify and resolve bottlenecks in labor utilization and improve response time.

Routine Processes: Are routine processes moving smoothly? Even if routine processes are running smoothly, are they as efficient as possible? Does the utility easily communicate what needs to be completed on a daily basis?

Planned vs. Reactive Activities: It is challenging to plan facility repairs and service in advance and thus personnel may currently be fixing problems as they arise rather than preventing them from happening. CMMS/AM software helps you proactively plan most maintenance tasks with standardized processes and set procedures. This will reduce reactive repairs and free up valuable time. The software enables movement from a reactive maintenance program to a preventative maintenance program.

Spare Parts: Have personnel ever failed to complete repairs in a timely manner because important spare parts were unavailable? Maybe the utility has the part but has no idea where it is currently located. Or personnel thought they had the part in stock only to realize it was used in a previous repair and another one was not ordered.

Waiting for parts is a common setback and source of inefficiency. CMMS/AM software can enable the creation of a spare parts inventory so it is possible to adjust new material orders based on stock level and identify patterns of materials consumption. This information allows the utility to gain control over its inventory and ensure critical parts are available when needed.

If the utility experiences some or all of the above issues, it is worth investigating CMMS/AM software further. The software is not inexpensive but is available at varying prices with varying capabilities and most systems will find a product that matches their needs. There must also be a commitment by staff to learn a new system; the payoff can make it worthwhile. The next step is to begin creating a list of what your utility wants out of CMMS/AM software before researching software companies.

It is very important to think about the utility's needs and create a list of features the CMMS/AM must have to help fulfill these needs. Split the list into essential features versus optional features. Many CMMS/AM software programs will do more than the utility needs and the utility may not necessarily want something that comes packed with thousands of features. A utility may be able to save money by knowing exactly what it wants out of the software. It is also important to not exclusively focus on the current needs; the utility should also consider what its needs may be in 5 to 10 years. The system does not want to outgrow its CMMS/AM software. The table of CMMS/AM criteria on the following pages will help the system select criteria that it wants in its software. It can be modified to fit the system's needs.

CMMS/AM software can be locally stored or cloud-based. Companies may only offer one or the other or both types of program location. There are advantages and disadvantages to either one. A cloud-based system requires fewer upfront costs because a utility does not have to purchase the storage device and their IT department does not have to set up and manage the server. A cloud-based system can also quickly implement updates and issues with the software can be solved remotely. An on-premises solution has better security and if a utility has data or regulatory requirements that force the data to stay on-site then this approach will be the best option. All CMMS/AM software should indicate whether the software is cloud-based or on-premises or both. When researching CMMS/AM software, utilities should ask their IT department if it has any requirements that mandate either cloud-based or on-premises solutions.

Hardware considerations for either solution include the acquisition of handheld devices to access the system in the field. Connectivity may be via cellular data, or it is possible to allow for offline data recording in the field with later syncing via wireless (Wi-Fi) service. Utilities should check with their IT department to determine if existing workstations are sufficient for handling access and usage with either solution.

Each CMMS/AM software will have a different pricing scheme. Pricing may be available on a company's website but more often it is necessary to request a price estimate. Most utilities will have to create a request for proposal (RFP) to purchase software. By creating an essential features list and researching CMMS/AM companies prior to RFP creation, a utility will be able to create an RFP that accurately identifies all of their needs while also staying within their budget.

It is important to understand the pricing structure and whether it includes an initial cost with annual maintenance fees (fairly common), only a one-time purchase price (uncommon), or a cost per year that is fairly uniform over time (sometimes). Almost all systems require annual fees of some type so it is important to budget for both initial and annual costs.

The ability of the software to meet the listed criteria and the cost of the software are not the only factors that should be taken into consideration. It is important to assess the stability of both the software and the company. Elements of this assessment include: the ability to handle computer operating software changes, including assurances that the software will be updated if an operating system change prevents the software from operating properly; years the company has been in business; number of clients served, especially the number of customers of a similar size and type; ability to export the data, including the types of formats that can be used; the customer service response times; and ability to receive customer service in person and/or remotely.

It is important to consider the training needs of the utility. What type of training personnel want, when the utility wants it, how the utility wants it delivered, and who should be involved should all be considered. It is important to determine if the price of training is included or is separate. Also determine what on-going training (past year one) might be desired and if so, what the cost of that training would be.

It is vitally important to schedule demos with the software companies being considered. The software needs to fit the culture of the organization and the skills and capabilities of the employees who will use it. It also needs to be user friendly, which can only be determined by the *actual users*. What is "user friendly" to one group of people may not be "user friendly" to another. Ask the companies to show the capabilities of the software in the areas of most interest. Schedule plenty of time for the interviews/demos and make sure that they are very willing to answer ALL of the staff's questions. If the company is not forthcoming with the type of answers the utility is looking for at the time of the demo, it is unlikely they will be forthcoming later. If not all the questions are answered, schedule an additional demo that focuses on the necessary additional information.

Ensure that the proposal review team contains a cross-section of employees and that they attend the demos of the software companies. Have each individual grade the proposals against the selection criteria and determine if there is a consensus conclusion on the best software for the utility. If there is a consensus, the utility should enter into negotiations with that software vendor. If there is not a consensus, it is important to discuss each person's views on the top contenders and try to reach a consensus.

The following represents a road map of the selection process.

8.1 Step 1: Form Your Software Selection Team

- The most important people to engage are those already familiar with maintenance and operations as well as facility managers.
- Include at least one person from each department that will be using the system, including billing and IT, to ensure that everyone's needs will be met.

8.2 Step 2: Understand Your CMMS Needs

- The more you know about your utility and its needs, the better prepared you will be to pick the right software. Knowing exactly what you want will also save you money.
- Ask yourself what you want out of CMMS/AM software. Do you want to be able to track work orders? Are you looking to do preventative maintenance? Do you want to be able to access information in the field?

8.3 Step 3: Understand Your IT Needs

- Do you want your CMMS/AM software to be cloud-based or installed on-premises? Companies will usually offer one or the other although sometimes both are possible.
- A cloud-based system offers fewer upfront costs. Additionally, upgrades and troubleshooting can be done remotely, saving more money.
- An on-premises solution provides better security. Check with your IT department to determine if your utility has requirements that force the data to stay onsite.
- Review the IT master plan, if one exists, regarding the potential to interface or integrate with other software used by the utility or larger organization the utility is within (e.g., financial management systems or human resource systems). If there is no IT master plan, use other options to determine the ability to integrate with other software.

8.4 Step 4: Create Your Essential and Optional Features Lists

- Once you understand your utility's needs, create an essential features list and an optional features list.
- It's important to not exclusively focus on your current needs. You should also consider what your needs may be in 5 to 10 years. You do not want to outgrow your software.

8.5 Step 5: Establish Your Method of Procurement

- Option 1: Your utility requires or desires a request for proposal (RFP) or similar process. Make sure the RFP complies with all state and local procurement rules and regulations.
  - If your utility is required to take bids or write an RFP, make sure to include all of the features on your essential and optional lists in the RFP. Also include whether or not the software must be cloud-based or installed on-premises.
  - Once bids come in, request demos of the top three to five candidates. You will not know if the software is right for your utility without demonstrations.
- Option 2: Your utility does *not* require an RFP or a similar process.
  - If your utility is allowed to pick software without an RFP process, begin researching companies using your features lists as parameters.
  - Once you have thoroughly researched companies and found some that check every feature off your essential features list, create a short list of companies, three to four, and set up a demo.

8.6 Step 6: Demo

- Companies will tailor the demo to reflect the items on your two lists.
- While the demo is going on think about functionality and ease of use. Does each feature seem user friendly?
- As each feature is shown, ask questions. For example, if preventative maintenance is on your list, make sure the software supports preventative maintenance in the way you do it or would prefer to have it done.

8.7 Step 7: Price

- Each CMMS/AM software will have a different pricing scheme. Pricing may be available on a company's website but more often you will have to request a price estimate. Do this early on to see if it is within your budget.
- Prices can be based on one or all of the following: gallons per day, number of connections, or number of users (concurrent or total). Have this information before asking for a price estimate.
- There are many hidden costs in CMMS/AM software. Below are questions you should ask the company to understand the full cost.
  - Is it subscription or license based? Does the license or subscription fee stay the same or increase every year?

  - How much are the onboarding and training fees?
  - Is there a cost to import and integrate data into the system?
  - Is there an additional cost for technical support?
  - Is there an additional cost for upgrades?
  - Are there additional modules? How much do they cost?
- Ensure cost information is handled appropriately according to the utility's and state's procurement rules/regulations. In some cases, the rules may require the price to be negotiated separately from the selection of the software when an RFP is used.

8.8 Step 8: Technical Support

- All companies should provide some type of technical support but the type may depend on whether your software is cloud-based or on-premises.
- The following questions will help you gauge the level of technical support provided.
  - How does technical support work? Is there around-the-clock support?
  - Do you provide instructional videos, in-software guides, live chat?
  - How are upgrades done? Are we shown the changes?

Chapter 4

Criticality

Criticality is the heart and soul of asset management. Understanding criticality enables resources to be put towards the greatest need for the utility, and expenditures on critical assets can reduce the overall criticality for the organization. Reducing criticality helps prevent or reduce the number and magnitude of catastrophic events that can provide financial, environmental, and social benefits. Using repair/replacement/rehabilitation approaches that are based on criticality can provide greater benefits for the amount spent or can allow an entity to spend less money to achieve the same benefit.

Two questions are important in determining the criticality of an individual asset:

- What is the probability or likelihood of failure of a given asset?
- What is the consequence if the asset does fail?

When considering the probability of failure (POF), there are four potential failure modes for an asset:

- Mortality
- Level of Service
- Capacity
- Financial Inefficiency

An asset can fail for any of these reasons, or it can fail for multiple reasons. Each of these failure modes is explained below.

Mortality Failures: A mortality failure is what often comes to mind when one thinks of a failure. It is the physical failure of the asset due to deterioration or destruction. It might be from a crack, a break, or something may have caused it to crush or be sheared off. It could be related to a disaster such as a fire, flood, earthquake, tornado, or a collision from a vehicle or other force. It can be anything that causes a physical issue to prevent the asset from performing its intended function. The asset may be repairable, but it will no longer function properly until it is repaired or replaced.

Level of Service Failures: A level of service failure is one in which the asset can continue to perform its function but not at the level needed to meet the requirements of customers. For example, the quantity of a contaminant in a well might rise to the point that it is above the regulatory standard. The well might still be able to produce water, but not below the regulatory requirements. A utility may decide to replace the well with a new one rather than adding treatment to the existing well.

Capacity Failures: A capacity failure is one in which the asset does not have the physical capacity or quantity required. For example, a pump that does not pump sufficient quantities of water, a well that does not produce enough water, a wastewater treatment unit that is undersized for the current flow.

Financial Inefficiency: Financial inefficiency failures occur when it costs more to continue to operate and maintain an asset than it would to replace the asset or the cost of continued operation and maintenance constitutes a large part of the overall asset cost. In these cases, the asset could continue to serve its function but not in an economical way. In some cases, this type of failure may occur if the asset is obsolete. Some assets, when kept in service for a long period of time, are no longer supported by the manufacturer. In these cases, it is not possible to purchase spare parts, so new parts have to be machine-made. It may be less expensive to purchase a new asset than to continue the maintenance.

It is important to note that the asset failure does not indicate the type of action that will be taken after the failure—repair, replacement, rehabilitation, or abandonment. That decision is part of the life cycle analysis of the asset. It also relates to the consequence of the failure, which is discussed later in this chapter.

The POF can be rated on a numeric scale in which the lowest likelihood of failure is the lowest number and the highest likelihood the highest number. When beginning an asset management program, a simple 1 to 5 scale, with 1 being an extremely low probability and 5 being a very high probability, is sufficient for most systems. The type of scale is shown on Table 4.1.

Larger, more complex organizations may wish to have scales that range from 1 to 10 or even higher than that. A system may also wish to start with

TABLE 4.1 POF Rating Scale

POF Rating	Description
1	Extremely unlikely to fail
2	Low chance of failure
3	Moderate chance of failure
4	High probability of failure
5	Extremely high probability of failure; asset already shows signs of failure; asset has failed

a smaller scale and increase it over time as they become more proficient with asset management. It is a matter of balancing the cost of adding complexity with the benefit of that increased complexity. When there is a large number of assets, it may be more beneficial to have a larger scale than with a smaller number of assets. The larger scale allows for more separation of probabilities of failure between assets.

When considering the consequence of failure (COF), there are three factors to consider:

- Financial
- Environmental
- Social

Consequences in all of these areas should be considered.

Financial: Financial factors include anything that is related to monetary considerations. These costs can include the costs of: the repair, downtime, legal issues, parts, contract services, or other assets damaged by the failure. These factors are generally relatively easy to monetize.

Environmental: Environmental factors include anything that relates to damage to the surrounding environment. Examples include: wastewater discharge into a water body, chlorinated water flowing into surface water, loss of habitat, or loss of aquatic life or wildlife. It can also include regulatory issues, such as violating a permit.

Social: Social factors include anything related to effects on people. The effects may be health and safety related to either staff or customers, inconvenience to customers, closures of businesses or schools, service disruptions or blocked streets. These consequences may also include loss of confidence in the utility or decreased satisfaction. In rare cases, people—either utility staff members or the general public—can be injured or even killed by an asset failure. In one extreme example, 13 miles of sewer pipe exploded in the city of Louisville, Kentucky, in 1981. Four people were injured in the explosions and considerable property damage occurred to structures and roadways. (McAlister & Federle, 2021)

Unlike financial factors, these latter two factors are not as easy to monetize. They may need to be considered in a qualitative, rather than a quantitative, way as a starting point. However, research has been ongoing into ways of monetizing these types of consequences. For example, The Water Research Foundation (WRF) has a publication, WRF 4451 titled "Managing Infrastructure Risk: The Consequences of Failure for Buried Assets" that developed an approach to monetizing the full triple bottom-line costs (financial, environmental, and social) (Raucher, 2020). Other researchers have been working on this task and have presented findings of their work at conferences. While the utility may wish to start more qualitatively at the earlier stages of implementing asset management practice, it should consider a more robust approach later that monetizes the environmental and social consequences. The literature on this subject can be used to determine an approach that best fits the utility.

Similar to the POF, consequence of failure is given a numeric score with the lowest number equating to the lowest consequence and the highest number the highest consequence. The rating scale is often the same for both POF and COF; if a 1 to 5 scale is used for POF, it is usually a 1 to 5 scale for COF. However, the scales do not have to be the same. Choosing a higher scale for POF than COF weights the probability of failure higher than the consequence of failure or choosing a higher scale for COF weights that factor higher. Further, the scale for both POF and COF can be higher than 1 to 5 if a system wants to have more gradation in its risk assessments. A 1 to 5 scale for consequence of failure is shown on Table 4.2.

A utility can take these scales and customize them to their own understanding of consequences. An example is shown in Table 4.3.

If a utility decides to make its own COF rating scale, such as the example above, it can decide how many factors to include in each aspect and how many need to occur to get that rating. For example, if there are serious

TABLE 4.2 COF Ratings

COF Rating	Description
1	Extremely low consequences
2	Low consequences
3	Moderate consequences
4	High consequences
5	Extremely high consequences

TABLE 4.3 Example of a Customized COF Rating Scale

COF Rating	Description
1	Financial: Less than $1,000 Environmental: No consequences Social: Fewer than 5 customers disrupted, no business disruption, no health or safety issues
2	Financial: Between $1,001–$5,000 Environmental: Less than ¼ acre of land affected, no waterways affected Social: Between 6 and 25 customers disrupted, 1 to 2 businesses disrupted, no health and safety issues
3	Financial: Between $5,001–$10,000 Environmental: Between ¼ acre and 1 acre of land affected, no waterways affected Social: Between 26 and 50 customers affected; 3–5 businesses disrupted; no health and safety issues
4	Financial: Between $10,001–$50,000 Environmental: Between 1 acre and 5 acres of land affected, minor waterway effect Social: Between 51 and 75 customers affected; 6–10 businesses disrupted; minor health and safety issues
5	Financial: More than $50,000 Environmental: More than 5 acres of land affected, serious waterway effects Social: More than 51 customers affected; more than 11 businesses disrupted; serious health and safety issues

waterway effects but no other environmental or social issues, would the COF still rate a 5? To address these types of concerns, the rating scale may say: "if any one of these factors is in existence the rating is a 5" or "if any two of the factors is in existence it rates a 5." A chart such as Table 4.3 can be as detailed as desired and have as many different conditions as desired.

A utility can also decide to create a more complex type of chart that rates each factor separately and then totals the factors to see where the probability or consequence ranks. An example for gravity wastewater collection pipe probability of failure is shown on Table 4.4. This table is intended for pipe that has not undergone any type of closed-circuit televising (CCTV) or other condition assessment techniques.

Each one of these factors can be given a score range based on the specific utility's experience. No two utilities would be expected to perform this type

TABLE 4.4 POF for Gravity Sewer Pipe

Pipe ID Number	Pipe Type Score	Pipe Diameter Score	Depth of Bury Score	Slope Score	Age Score	Soil Type	Known Failure Score	Total
11111	3	3	2	1	3	3	2	17
22222	2	1	2	1	3	2	2	13
33333	4	2	4	2	2	2	1	17
44444	5	4	1	2	1	4	4	21
55555	2	5	5	4	4	3	1	24

of analysis the same way because they are highly likely to have different types of failures. In one system, vitrified clay pipe might be the worst pipe type and in another, it might be the best. In some systems, there may be little to no observed difference related to pipe diameter while in others that might be a very significant factor.

In the example presented in Table 4.4, each factor may be given a specific score ranking based on identifiable factors. This approach allows for a more objective rating system. An example for pipe type is shown on Table 4.5.

In this example, the utility experiences the most failures with Vitrified Clay Pipe and the least with polyvinyl chloride (PVC). A table similar to 4.5 should be created for each factor included in Table 4.4 so each pipe can be given a ranking. If any factor is unknown, which is likely the case with pipe, a score can be included for unknown. For pipe type, it is likely that unknown pipe types occur more for older pipe than newer pipe, so a score of 3 or 4 for unknown may be appropriate. For other factors, unknown may be ranked high or low depending on the utility's experience.

In the same way, consequence of failure can be ranked. Table 4.6 shows an example of consequence ranking for gravity sewer pipe.

Similar to the case with the POF scores, tables can be created to determine the scores that should be given to each pipe for consequence of failure. An example with proximity to water body is shown in Table 4.7.

TABLE 4.5 Example of Type of Gravity Sewer Pipe and POF Scores

Type of Pipe	POF Ranking to Enter Into Table 4.4
Ductile Iron	2
Cast Iron	4
Vitrified Clay	5
PVC	1
HDPE	3

TABLE 4.6 COF for Gravity Sewer Pipe

Pipe ID Number	Proximity to Water Body Score	Critical Customer Served Score	Population Served by Line Score	Potential to Cause Health and Safety Issues Score	Proximity to Businesses Score	Total
11111	1	1	4	3	4	13
22222	2	1	2	4	4	13
33333	2	5	1	2	1	11
44444	3	4	5	3	3	18
55555	4	3	2	5	3	17

TABLE 4.7 Example of Gravity Sewer Pipe and COF Scores

Proximity to Water Body	COF Ranking to Enter Into Table 4.6
Within 50 feet	5
Between 51 and 100 feet	4
Between 101 feet and 500 feet	3
Between 501 feet and 2000 feet	2
Greater than 2000 feet	1

In the case of pipe, this kind of risk analysis can be used to prioritize pipe for CCTV work, since more information would be beneficial before choosing to replace pipe. It is important to note that risk analyses can be used for many purposes, not just determining asset replacement. A further discussion of the calculation of risk using this approach will be provided in the next section. It is important to remember that the simple approach of 1 to 5 for POF and COF can represent a good starting point and may be all that is needed for small to medium-sized utilities.

1.0 CALCULATING CRITICALITY

To calculate criticality, it is necessary to multiply the POF score times the COF score. If a 1 to 5 scale is used for both POF and COF, the criticality will range from 1 to 25. If a 1 to 10 score is used, the criticality will range from 1 to 100. The calculation is shown below.

POF × COF = Criticality

While some systems have used decimals to give POF or COF scores less than 1, it is not recommended that a score of 0 be used for either POF or COF. Using 0 for either POF or COF would render the total criticality score a 0 due to the multiplication. Every asset has at least a small chance of failure and a small potential for impact so it is necessary to have criticality scores above 0 and to use POF and COF numbers greater than 0. When decimals are used, it is a way to give POF and COF scores between 1 and 5 but have more gradations. For example, 1.1, 2.3, 4.4 can be used. This may be unnecessarily complicated for a small system but may give a larger system more nuanced scores.

If the more complex version of criticality is used, as shown in Tables 4.4 and 4.6, the criticality calculation is slightly different. The scores of POF and COF can be multiplied to get the total score and then the final result can be normalized to a standard score of 1 to 25 or 1 to 100 or some other range. The example data with normalized values from 1 to 25 and 1 to 100 is shown in Table 4.8. The normalized values put the numbers into a more manageable range.

2.0 WAYS TO REDUCE RISK

To reduce risk, actions can be taken to reduce the POF, the COF, or both. These actions can move an asset from a higher risk box to a lower risk box. Examples of actions to reduce risk are provided below.

- Routine and preventative maintenance (reduces POF): Performing the proper maintenance on the appropriate schedule for assets can help reduce the potential for failure. This strategy has limitations given the fact that assets can fail in ways unrelated to maintenance items (e.g., unexpected failures, discussed in life cycle costing) and maintenance can be performed incorrectly or on the wrong schedule which may be detrimental to an asset.

TABLE 4.8 Criticality Calculation

Pipe ID Number	Total POF Score	Total COF Score	Criticality Score	Normalized Score 1 to 25	Normalized Score 1 to 100
11111	17	13	221	6	23
22222	13	13	169	1	1
33333	17	11	187	3	8
44444	21	18	378	22	88
55555	24	17	408	25	100

- Monitoring (POF): Assets that have severe consequences of failure can be monitored to determine the current condition, the rate of asset deterioration, and any signs of potential failure. Monitoring is likely to involve some level of technology and possibly outside consultants so it can be an expensive strategy. Therefore, monitoring may be reserved for those assets that are deemed to be moderate to high risk of failure and moderate to high consequence of failure. The monitoring can allow a utility to intervene before the asset failure to reduce the potential for the failure to happen. Examples of monitoring can include a storage tank inspection with a diver or robot, pipe inspection with CCTV, vibration analysis, lubrication analysis, or thermal analysis. A more recent innovation is real-time pipe monitoring using fiber optic cables as a sensor. This technology would be deployed in highly critical pipes to allow real-time monitoring of the pipe. At this point, it is most likely only cost-effective for large utilities, but over time, as the technology increases in adoption, it may be able to be deployed in other utilities.

- Spare Parts (POF & COF): Maintaining an adequate supply of spare parts for consequential assets can ensure that the asset is up and running quickly after the failure occurs, reducing consequences related to an asset being out of service for a considerable period of time. Having spare parts available also makes it possible to replace parts that are showing signs of wear before the asset completely fails, thus reducing POF. To ensure this strategy actually reduces POF and COF, it is necessary to have an organized spare parts inventory, knowledge of the parts that need to be replaced, and storage facilities that will ensure the parts are kept in a suitable environment (not too hot or cold or subject to sun or other weather conditions). The parts should be easy and quick to find when needed in an emergency. Because spare parts can be expensive, it may be necessary to make decisions regarding which parts and how many to keep on hand. Priority should be given to those assets whose failure will cause severe consequences (whether high or moderate probability of failure) and those assets with moderate consequences but high probability of failure.

- Specialized Training (POF & COF): Some assets may be harder to maintain, fix, or operate than others. Having specialized training for staff to ensure that they are operated correctly, within their design limits and according to specifications, as well as the manner to maintain the assets properly can help ensure that the assets do not fail prematurely and that they can be addressed quickly to reduce consequences from the event.

- Replace Assets Early (POF & COF): There are some assets for which a failure cannot be tolerated; the consequences are so severe the event cannot be allowed to happen. An example of such an asset could be a high-producing well, especially if it is the only well for the entire community. In this case, if the well fails, it may not be possible to provide any water to the community and water may have to be trucked into the town at considerable expense. A well takes a long time to drill and bring online, so the situation could be extremely dire for the community. To prevent this occurrence, a new well should be installed *before* the old well shows significant signs of failure. It may take 3 years or more from start to finish to put the new well in place, including planning, design, construction, permitting, and securing the necessary funding so this strategy should be combined with a monitoring strategy. If the well is monitored on an annual basis, signs of deterioration can be noticed soon enough to allow for the lengthy replacement process. This strategy gives up a portion of the asset's life in favor of reducing risk so it should not be used for low-consequence assets. In those cases, the assets can follow a "managed failure" strategy—meaning that the event can be allowed to occur and can be dealt with adequately after the failure.
- Redundancy (COF): In the case of redundancy, multiple assets are installed to do the job, such that not all assets are needed at once. It is not redundancy if two pumps are installed and both are needed during high flows. In that case, it would be necessary to have at least three pumps so one of the pumps is available as a standby pump while the other two are working. Installing additional pumps beyond the two required provides increased redundancy. If the redundant asset can immediately start working to ensure continued service, the bad consequences do not occur. As an example, a sewer pump station may have an emergency generator to provide redundant power. If the power goes off, the generator can immediately start operating to provide power to the station. Other examples are multiple wells, multiple valves, multiple fire hydrants, looped drinking water pipelines, multiple storage tanks, or multiple treatment trains. Some redundant assets are portable, such that they reduce risk but not to the same level as those that can be immediately put into service. Portable assets may include portable pumps or generators. The problem with portable units, such as an emergency generator, is that they can only be used at one place so if a condition occurs that cuts power to multiple facilities all at once, some facilities will still fail. While redundancy is an excellent strategy to reduce risk, it can be costly because it requires more assets than are strictly needed to do the job. Also, redundancy

is not foolproof. The same conditions that caused one or more assets to fail may cause even the redundant asset to fail. For example, if the pump station was struck by lightning and the controls were knocked out, all the pumps could experience the same failure and none of them may work properly.

While redundancy is primarily a strategy addressing COF because the operation of the redundant asset prevents the loss of service, it may have some positive effect on the life of the asset that keeps the POF score lower for a longer time span. While the POF score is not going to be reduced from what is assessed to be the likelihood of failure at that point in time, it may be the case that the asset lasted much longer overall that would have been expected. For this type of occurrence, the redundant asset needs to be rotated in service to allow the main assets to work fewer hours overall, thus reducing somewhat the stress on the asset and lengthening the time before the asset failure occurs. How much of an effect this has on the asset depends on the amount of redundancy and the operational strategy used. Some redundant assets will not experience this reduction, such as multiple isolation valves in a drinking water system. The presence of valves further up the pipe line means that the line can be shut off even if the closest valve fails, but the presence of these other valves does not have any effect on the probability of failure of the isolation valve. Other examples are looped water lines and fire hydrants.

It is not possible to implement each of these strategies for every asset, given the resource constraints on the organization. However, the high-risk assets can be examined to see which strategies make financial sense and those strategies should be implemented to reduce overall risk.

Redundancy: There are a couple of ways to consider redundancy in the calculation of risk. It can be considered when the POF and COF are being evaluated for that particular asset. Based on the amount of redundancy, the COF can be reduced from a higher number to a lower number to demonstrate the lower likelihood of consequences. The POF can also be reduced if the operational strategy is such that the probability of failure decreases based on the presence of other assets. As an example, consider a pump that is showing signs of failure, has not been well maintained, and is in fair condition. This pump may receive a 4 rating for consequence of failure. If it has minimal redundancy, it may be dropped to a 3 rating. If it has extensive redundancy, it may drop to a 2 or possibly even a 1.

Another way to include redundancy in the analysis is to apply a redundancy factor to the calculation of risk. In this case, the redundancy factor is a score above 0 but less than or equal to 1. The redundancy factor cannot

TABLE 4.9 Redundancy Factors

Level of Redundancy	Redundancy Factor
0%	1
33%	0.875
50%	0.75
100%	0.5
200%	0.25
300%	0.125

be 0 because that would allow the overall risk to fall to 0 and that is never the case. There is always a non-zero risk that an asset will fail and that it could cause at least some consequence, even if very minimal. If something could fail and have absolutely no effect on the operation (zero risk) it is likely not an asset to begin with, but rather a consumable or a supply.

The redundancy factor is 1 if there is no redundancy; the risk of the asset does not change. The redundancy factor is lower the more redundancy there is. A utility can define its own redundancy factors based on the level of redundancy. As an example, some potential redundancy factors are shown in Table 4.9.

There are many other possible factors based on level of redundancy as well. The utility is free to choose the strategy that makes the most sense for them. If redundancy has proven to be a good strategy to reduce risk, factors may be lower (closer to 0); if redundancy has not been a very successful strategy, the factors should be higher (closer to 1). However, redundancy should never be reduced to 0. That would mean that the utility believed that redundancy took all risk away. This is not the case, because whatever caused the asset to fail may cause failure of all of the redundant assets. Consider one prominent example: the Fukushima Daiichi nuclear disaster after the earthquake and tsunami. In that case, there was redundant power and other ways to dissipate heat, but all failed due to flooding by the tsunami (World Nuclear Association, 2022).

The equation becomes the following if a redundancy factor is used:

$$\text{POF} \times \text{COF} \times \text{RF} = \text{Criticality}$$

If an asset has a POF score of 4 and a consequence of 3 with no redundancy factor used, the risk score would be 4 × 3 or 12. If a redundancy factor of 0.75 were applied, the risk would drop to 9. If on the other hand the redundancy was included in the decision of what COF rating to give, the rating might be reduced from a 3 to a 2. The criticality score in that case would be 4 × 2 or 8. This score is similar to the application of the redundancy factor (RF), but that will not always be the case. The use of the redundancy factor may have more effect than reducing the criticality score depending on the level of redundancy.

It is important to remember that both ways of addressing redundancy cannot be used at the same time. Either the RF factor is used or the COF score applied to the asset is reduced, but not both. If both are done, the criticality will be lower than it should be.

3.0 VISUALIZING RISK

It can be helpful to visualize risk in some way through a simple, easy-to-see graphic. This visualization can be done in different ways depending on the number of assets and the information that is being conveyed. It can be done on the system as a whole (i.e., every asset in the system), on a subset of assets of different kinds (e.g., all the assets in a pump or lift station) or just one asset class (e.g., pumps, tanks, valves, manholes). Some examples are shown below on Figures 4.1, 4.2, and 4.3, but there are many other ways to demonstrate the criticality results.

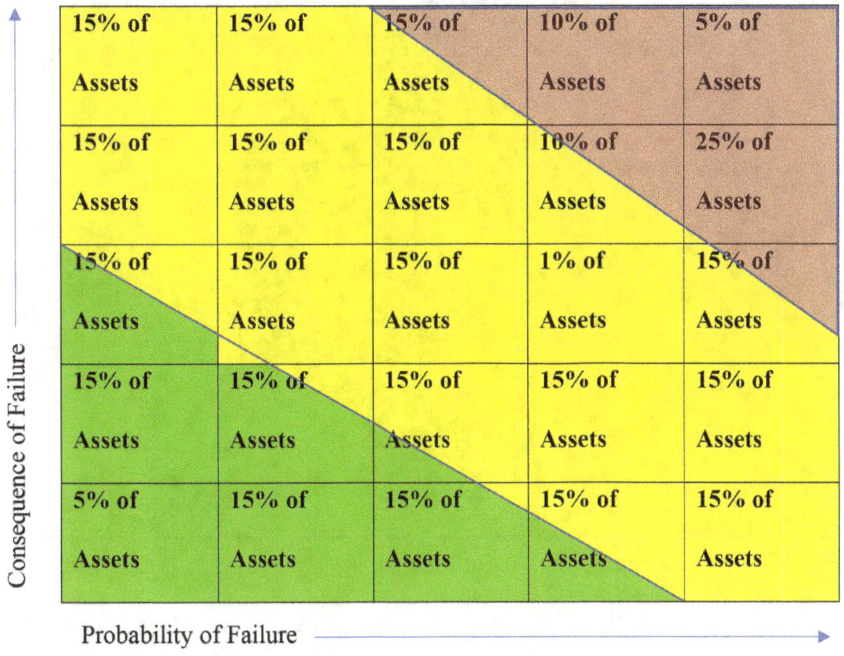

FIGURE 4.1 Risk Visualization Example 1

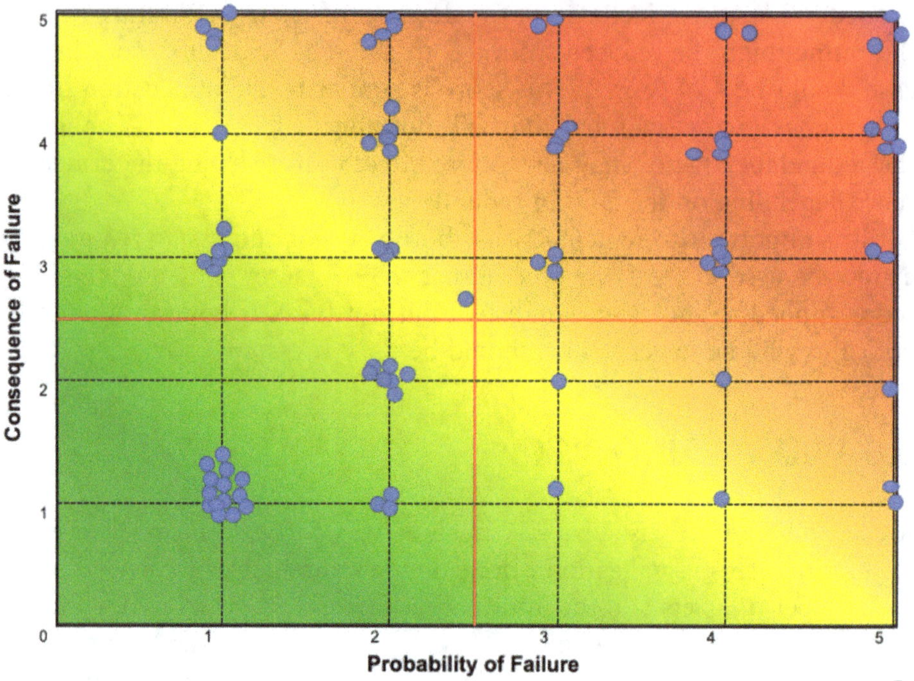

FIGURE 4.2 Risk Visualization Example 2

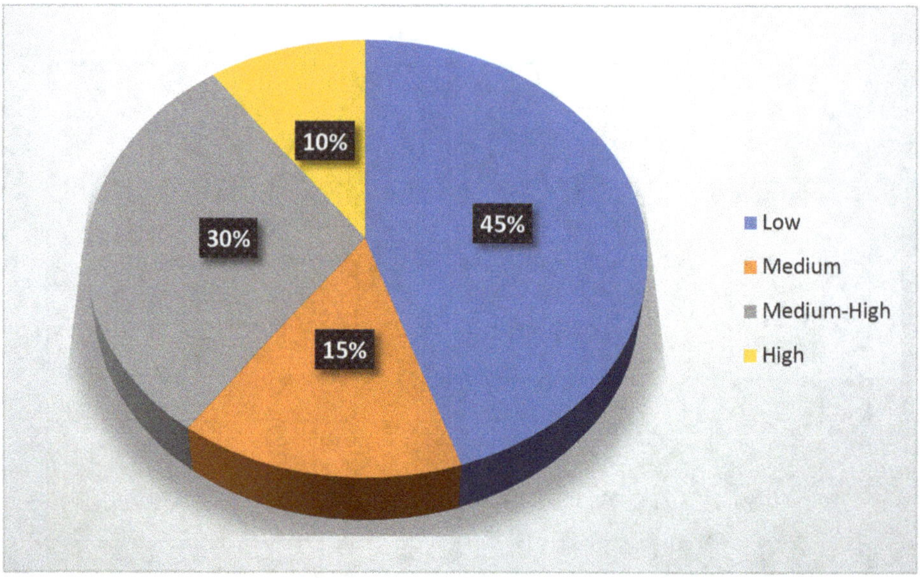

FIGURE 4.3 Risk Visualization Example 3

There are other ways to visualize the data and utilities should choose the method that works best for them. If the system has few assets, the actual name of the assets can be used in the visualizations. If the system has more assets, the number or percentage of assets will probably need to be used. A list of the assets that match up to the criticality number or ranking (e.g., low, medium, high) can be created and used as an attachment to the visualization. Also, more than one way can be used to visualize assets. Assets can be visualized as a class of assets (e.g., pumps, tanks) or can be visualized based on a location (e.g., Westside Treatment Plant, Palm Street Pump Station).

4.0 RISK UNDERSTANDING/RISK TOLERANCE

It is important for elected leaders/decision-makers to understand the risk level of the current assets and the overall facility and take that into account as they make decisions about current and future investments in operations, maintenance, assessments/inspections, repairs, and rehabilitation, and potential replacements. They should also understand that choosing the level of investment relates to their tolerance for risk. Investing a lot of money to reduce risk prevents or reduces the occurrence of catastrophic failure, while investing little may mean that serious consequences will occur.

The risk of negative consequences must be balanced against the amount of investment needed to reduce the risks. It may take time to move the organization from a higher risk setting to a lower one. It is difficult to remove all risk from a drinking water, wastewater, or stormwater utility because there are events outside the control of the utility, unexpected asset failures can occur, and there is insufficient money to replace every asset when it starts to show signs of failure. However, working towards reducing the overall risk to the extent possible can improve the overall operations and management and reduce the long-term costs associated with the facility.

As an example, consider the risk profile shown in Figure 4.4. There are assets in all four risk quadrants, but quite a few are in the high-risk quadrant (red zone on the chart). The utility would need to invest additional funds, in some way, to decrease the risk. Those funds may be for additional staff, equipment, supplies, repairs, rehabilitation, and/or replacements. Without investment, and probably significant investment, the elected leaders/decision-makers would be tolerating a level of risk that would allow at least some number of serious consequences to occur.

Similar charts could be prepared for individual asset classes to explore the risk associated with different types of investment decisions. For example, isolation valves could be reviewed to determine how many would cause serious consequences if failure occurred. Then a decision could be made to invest sufficient funds to reduce the risk associated with valves.

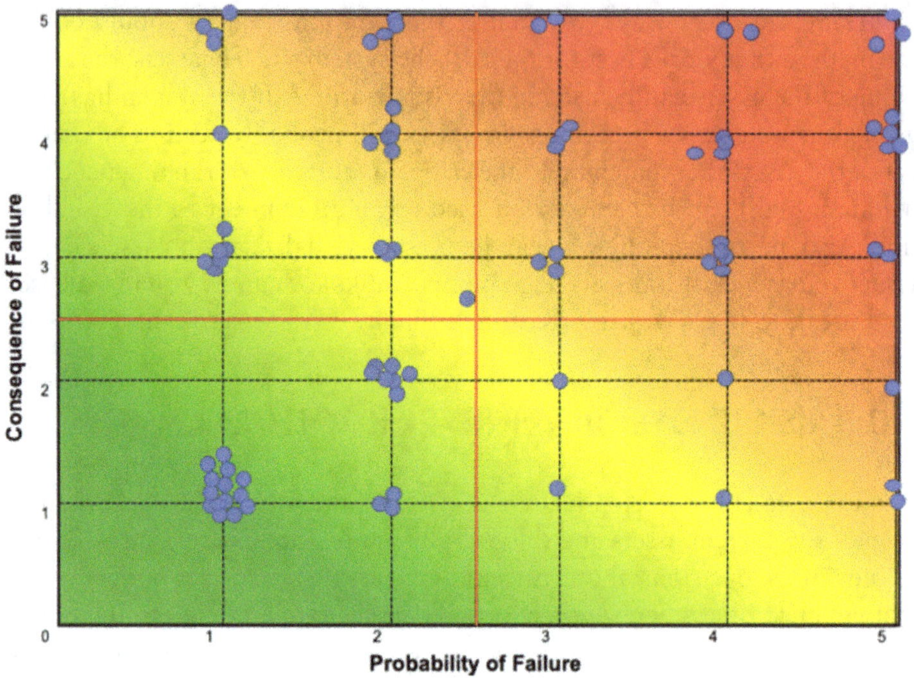

FIGURE 4.4 Risk Understanding/Risk Tolerance Example

Over time, a utility can lower its risk by investing in asset maintenance, rehabilitation, and replacement to ensure that there are few assets in the highest risk category. This will help reduce the number of bad incidents but may also increase the overall budget, at least for a period of time while the utility is moving from higher to lower risk. A utility should examine its overall risk and determine whether their current risk fits their willingness to tolerate the risk of serious or catastrophic consequences. If not, the utility should invest strategically to reduce risk to the level of tolerance.

It may not be necessary in all cases to increase the investment level to achieve the desired results. In some cases, a change in *how* the utility spends its money can lead to better results with no increased investment. If the utility has not been following a risk-based decision-making process, it is likely spending money on low- or medium-risk assets along with high-risk assets so it will not achieve the full reduction in risk that can occur with more targeted spending. An example demonstrating this concept is described below.

5.0 CRITICALITY EXAMPLE

A large utility wanted to examine its water main breaks in small diameter pipes in the distribution system (less than 12 inches in diameter) and determine if there was a better way to address repair and replacement decisions. Paper-based work order data was available for 15 years of repairs on mains up to that point. The data showed that from year to year, the amount of investment in small diameter replacements was relatively similar, ranging between $1 and $2 million per year and the number of breaks was relatively stable at approximately 300 per year. Clearly, the 15 years of investment had not made a considerable difference in the outcome and the utility was looking for a better way to determine pipe replacement decisions. They embarked on an asset management journey and wished to apply asset management principles to this issue.

A detailed analysis of break data showed the pipe breakage issue was primarily focused on one pipe type that was creating half the breaks but accounted for only 5% of the total system pipe. The utility knew this pipe type was a problem but during the 15-year period, had been choosing replacements primarily based on ease of construction. They were choosing contiguous sections to replace (e.g., replace 1 mile at a time, rather than small sections here and there) that contained a problematic area and were working their way from one side of the system to the other in replacing that type of pipe. The problem with the strategy was that many good sections of pipe were being replaced along with some bad sections and many terrible sections had to be left in due to the limited funds available for replacement. There were approximately 80 miles of pipe needing to be replaced and at the current rate, it could take 40 years or more to make the replacements. If the wrong decisions were made, many streets would need to be repeatedly dug up to repair pipe while it was waiting to be replaced. This approach angered residents who saw multiple pavement cuts in their area, sometimes on relatively newly repaved streets.

The asset management approach opened up a new way to consider pipe replacement. Each pipe segment was defined as an asset and evaluated for its probability of failure and its consequence of failure. Then an overall risk could be applied to each segment. When the segments were plotted on a map, with red, yellow, and green coloring to indicate high, medium, and low risk, respectively, the approach became much clearer. There were small segments of pipe all over the distribution system that should be replaced first before any other replacement was undertaken. This approach meant that more mobilization was required during construction and therefore, less pipe overall would be replaced. This idea was initially met with disagreement from the engineers on staff, but the utility pressed forward. Within 5 years of switching to an approach that took out the worst sections of pipe and left the better sections in place, the utility saw a 45% reduction in break rate and a cost avoidance savings of approximately $9 million dollars (over the 5 years). The utility had kept the overall investment in this pipe replacement program similar during the time period.

The reason for the dramatic change was related to criticality. Initially, the utility had been replacing pipe that was a mix of low risk, moderate risk, and high risk. With the change, the utility was only replacing high-risk pieces of pipe. Each year, fewer high-risk pieces of pipe remained in the system, reducing the number of breaks that occurred. In this case, the same amount of investment, spent differently, enabled a much greater effect.

6.0 REFERENCES

McAlister, S. & Federle, J. (2021). Where were you when the sewers blew? Louisville marks 40 years since sewer explosions. *WHAS11 Louisville News*. https://www.whas11.com/article/news/local/the-vault/where-were-you-when-the-sewers-blew-marking-40-years-since-lou-sewer-explosions/417-72c8 4696-fe89-4fc1-8a71-015d940f0c3d

Raucher, R. (2020). *Asset management framework for forested and natural assets* (Project 4727). Water Research Foundation.

World Nuclear Association. (2022, May). *Fukushima Daiichi accident*. https://world-nuclear.org/information-library/safety-and-security/safety-of-plants/fukushima-daiichi-accident.aspx

Chapter 5

Life Cycle Costing

Life cycle costing is probably the most involved and complicated portion of asset management. It combines many different elements to enable decision-making from the start of the asset's life all the way through the end of life. Life cycle costing represents the intersection of managing individual assets in a cost-effective manner (managing assets) with strategically managing the entire collection of assets (asset management). There must be an understanding of the full life cycle of the assets as well as the range of interventions that are possible during each stage of the life cycle. Because each stage can include a wide array of actions and it is not possible to do every action on every asset at each stage of its life, it is necessary to prioritize which actions are taken, when, and how. Those actions that are likely to have the greatest net positive effect to uptime (the amount of time the asset remains in operation without an issue) and/or the useful life of the asset (extending the amount of time the asset can remain in service doing the job for which it was intended) are the most beneficial to the organization.

This portion of asset management also requires the collection of data in order to make informed decisions. Therefore, having a place to store this information and being able to retrieve it will become valuable. The greater the ability to analyze data, the more opportunity to improve decision-making around all aspects of managing assets and asset management.

An additional component of life cycle costing is understanding that all costs of the asset's life cycle should be considered when deciding on the best alternative to undertake when assets fail or new projects are contemplated. Considering only the initial cost of the project, but not future operation and maintenance, repair and rehabilitation, spare parts, potential staffing, and disposal, may mean that the decision-makers choose the incorrect option and one that leads to greater costs overall or less efficient operation. All aspects of the life cycle, including costs, should be considered when making decisions, especially when making decisions about high-dollar projects.

1.0 LIFE CYCLE OF AN ASSET

A fundamental part of life cycle costing is understanding the components of the life cycle of the asset itself. The life cycle starts long before the asset is installed and the "keys are turned over." The asset's life cycle includes the following components as shown in Figure 5.1.

The asset's life starts in planning or what is sometimes referred to as conceptual design. The next stages are detailed design, construction, operation, maintenance, repair/rehabilitation, replacement, and finally disposal or abandonment. There may be a few assets for which one or more of these components are not relevant (e.g., an asset that cannot be repaired or rehabilitated but only replaced) but most assets will include all of them. The importance of each component may also vary based on the asset; complex assets will require careful consideration during construction, while mechanical assets should receive the appropriate maintenance. Some assets will need to be disposed of in specific types of landfills, while others can be abandoned in place, and still others can be recycled.

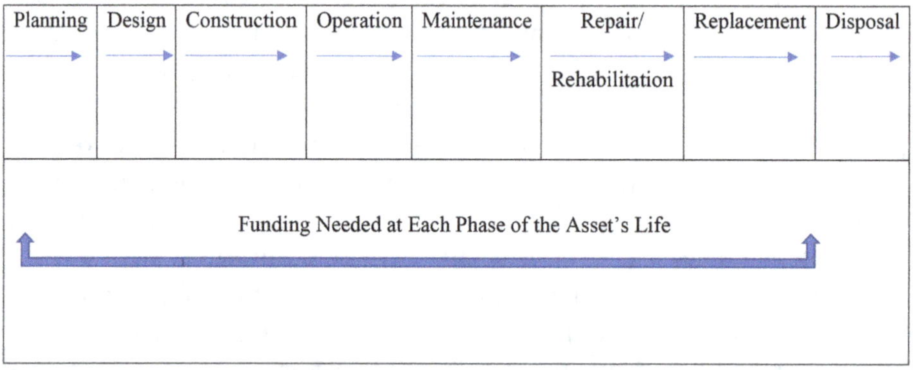

FIGURE 5.1 Asset Life Cycle

A description of the asset life cycle phases is contained below. Following each description is a list of potential interventions that can be conducted during that phase. Not every intervention of this type would be completed for each asset or project and there are other interventions possible during the stage. The interventions are intended to improve decision-making, extend the asset's life or increase its uptime.

These descriptions refer to an "asset" but it is highly likely that the utility will be completing activities that involve many assets. The same life cycle exists whether it is one or many assets. One asset is being described here for simplicity of describing the life cycle but it should be understood that the same life cycle mindset can be applied to collections of assets or complex projects that involve a considerable number of assets.

Planning: The asset's life starts when it is first considered in initial planning or conceptual design. At this point in the asset's life, it is just in initial consideration and all options are available. It can be any size, type, material, location, and any changes are very low cost or no cost to make. The asset is just "on paper" or in someone's mind so alternatives can be easily discussed and reviewed. This is an excellent time to bring the operations and maintenance personnel into the discussion to ensure that their perspectives are considered and their concerns addressed.

Furthermore, the results of previous asset management activities should be brought into planning. For example, the projects being considered for planning should be on the utility's capital improvement plan created by the asset management program. Bringing the results of prior asset management work into planning creates the cycle or loop that allows continuous learning and improvement. This approach incorporates the idea of the Deming Cycle—Plan, Do, Study, Act—that it intended to foster continuous improvement (Deming Institute, 2025). The planning portion of the life cycle is the first part of the Deming Cycle, while the Do, Study, and Act phases will come later.

> *Potential Interventions:* Assessment of overall need for the project or asset(s); consideration of the overall benefits and the risk of not doing the project; consideration of non-asset solutions; alternatives evaluation; consideration of potential operation and maintenance requirements of each option; consideration of the location of assets; analysis of expected total life cycle costs of each alternative; consideration of results of asset management activities completed up to this point; change in type of design; or change in types of assets, brand, or manufacturer.

Design: The second phase of the asset's life is design. In this phase, the initial planning moves to more detailed design. At the beginning of this phase, there is still a fair amount of ability to make changes, such as changing locations or even choosing a different alternative. As the design phase moves along and it progresses from 30% complete to 60% to 90% complete, it becomes harder and more expensive to make any changes. By the time the end of this life cycle phase is reached, and the project design is 100% complete, it is difficult to make changes and if changes were made, it would be very costly. However, the project is still "on paper" at this point, so it is less expensive to make changes here than if the project was in the construction or operational phase. If the utility felt that it should no longer proceed with the selected alternative or needs to make some other kind of change, it is better to do so at this point (less expensive and easier to accomplish) than at later stages when the project is physically being built or is in operation.

Operation and maintenance personnel should be given the opportunity to engage with design staff in reviewing and offering up design recommendations that facilitate proper maintenance (e.g., layout and work space considerations; monitoring and maintenance equipment add-ons, such as limit, level, flow or pressure devices, or lifting equipment for machinery; ancillary equipment and/or spare parts requirements; material and coating preferences).

> *Potential Interventions:* Considerations of the chosen option; materials of construction; manufacturer of assets (may have limited ability to specify a particular manufacturer due to procurement rules); matching the project/assets to the existing system; designing out known points of failure; specific location of assets; considerations of operability; reviewing utility's past history with design of this type of asset(s); or resolution of any known design issues.

Construction: After the project is designed, it moves into construction. At this point, the asset's life from this point forward is mostly locked in. A change during construction likely requires a change order which can be very expensive. Understanding the asset's performance over time can lead to changes in future construction practices. It is important to track causes of asset failures that are related to construction practices. While there may be little that can be done to address the assets already in the ground, future asset replacements can be done using different, and more appropriate, construction techniques.

> *Potential Interventions:* Specifying construction techniques to improve asset performance or asset life; addressing construction concerns as they occur; fixing observed problems/incorrect design;

pictures of assets for the asset inventory; gathering asset attributes for asset inventory; accepting asset inventory information from contractors; or documenting exact location of assets.

Construction oversight is an important intervention. The utility may have a construction inspector on site all the time or most of the time to represent the utility's interests and ensure the construction is completed according to design specifications. However, operation and maintenance personnel should be given the opportunity to observe some of the construction process to see how assets were installed as this may improve future repair of assets, particularly buried assets. They can also observe whether there are incorrect procedures or problematic techniques being used. If such observations occur, changes should be considered, even if it costs the utility money in the form of a change order. It is much better to address the problem right away rather than dealing with it for many years to come. It is likely that the cost of the change order is far less than future expenditures related to repairs. Operation and maintenance personnel should also be provided an opportunity to review submissions of operation and maintenance data, such as manuals, manufacturer certifications of proper installation and start up, asset data, and performance testing results.

After construction is substantially complete, the utility is ready to accept the assets and begin operations. It is important that the addition of the assets into the asset management system or computerized maintenance management systems (CMMS) system occurs before the operation begins so that the utility can document maintenance records, asset failures, inspections, and other tasks during the initial warranty period(s). Also, any data needed from outside vendors should be secured while the vendors are still under contract, as it may be much more difficult to obtain this information after they leave. The utility should specify clearly the manner and format in which the data is to be provided to the utility.

Operation: By the time the asset is put into operation, much of the asset's costs are locked in. This situation is the reason that an asset's life must start earlier in the process. The influence is gained sooner and must be exerted before the asset is chosen and installed. While operation is often linked with maintenance in what is generally called "operation and maintenance," it is its own distinct phase of the life cycle that includes strategies for the cost-effective, efficient, day-to-day running of the facility.

In this phase, the utility will want to consider the day-to-day running of assets, such as: managing lead/lag pump operation, adjusting storage tank or wet well level operation to ensure pumps are not cycled too frequently, determining whether the operation of the asset should be electronic or manual, and determining when individuals will be physically located at the plant. In some cases, regulations may stipulate when operators need to be present, such as when water is being produced, while in other cases, the utility may have more latitude regarding when the facility is staffed (e.g., only during daylight hours, just Monday through Friday, once per week).

This is also the phase in which to consider how emergency operations will be handled and how it might differ from routine operations. It is necessary to understand which assets are required during emergency operations; whether these assets are routinely checked to ensure they will operate when needed; how operator on-call is handled, including, who, when, how contacted; and whether there is a backup person from a neighboring utility (or elsewhere in the surrounding area) in cases where all operators are unavailable.

Potential Interventions: One of the most important interventions in this phase of the asset's life is to create an operation plan that is accessible to every employee who could potentially need it as well as any outside operator who may come in to operate the plant during a severe emergency in which the normal operators are not available. The plan should be easily accessible to all who need it, including any outside personnel.

Another potential intervention is to examine current operations to determine if activities are optimal for extending the life of the asset. For example, what part of operation causes the most stress on the asset? Is it when the asset starts? When it stops? While it is running? Will the starting or stopping of the asset have the potential to cause damage to other assets, as is the case with water hammer? Operational strategies that can increase the uptime (the amount of time the asset is running without a failure), lengthen the asset's life, or prevent damage to other assets should be employed whenever possible.

It is important that employees or outside contractors or vendors who will touch the assets understand how to operate them. This might require specialized training to ensure assets are operated properly. As one example, some utilities require fire department personnel or volunteer firefighters to undergo training in how to properly operate a fire hydrant, especially how to properly close a hydrant. Similarly, utilities may require training for construction companies or others who wish to take water from hydrants. In some cases, these entities are required to be certified every year in how to operate the hydrant. The certification comes from the utility after they pass a test on proper operation.

Maintenance: An asset undergoes maintenance during this phase of its life. The maintenance may be routine, preventive, predictive, or corrective. The most important part of maintenance is that the right maintenance, performed the right way, at the right time is practiced. Improper maintenance can be more harmful in some cases than no maintenance at all. The old adage, "if it ain't broke don't fix it," can be true in some cases. Similarly, more maintenance may not be better. As an analogy, if two aspirin cure a headache, you don't need to take four. The extra two pills either will be excreted from the body with no effect other than wasted pills and wasted money or the extra pills can cause harm. It is important to ensure that the manufacturer's recommendations as well as the system's actual experience with the assets drives action.

> *Potential Interventions:* Track maintenance on assets against the actual performance of the assets to see if the maintenance is having an effect. For example, if lubrication is currently on a schedule of once per year, but the utility notices that problems are occurring in the assets due to inadequate lubrication, the schedule may need to be increased.
>
> Another intervention related to maintenance is monitoring to understand the effect of the actions. Using the same lubrication example, consider the case of assets being lubricated four times per year. The utility wonders if this is too often and does lubrication testing at the end of quarter. The analysis shows no breakdown of the lubrication and no signs of excessive wear. At that point the utility could back off the lubrication schedule and continue monitoring until it finds an appropriate schedule that does not waste lubricant or cause excessive wear on the equipment.
>
> Training is an important intervention in maintenance. Personnel must understand how to do proper maintenance including what tools to use and the proper way to use them.

Repair/Rehabilitation: During an asset's life, it is likely to eventually require either repair or rehabilitation. Each time the asset fails, or shows signs of a future failure, a decision has to be made regarding how that failure, or potential failure, will be addressed. In cases where the decision is made to keep the asset in service, it will need to be repaired or rehabilitated to continue to provide the intended service. Repair is an action to fix only the broken element(s) of the asset but keep the rest of the asset intact. Rehabilitation is when more elements of the asset being addressed in order to restore the asset to as close to its original condition as possible, without replacement. An example of a repair on a piece of pipe would be placing a clamp around the break or splicing in a short piece of new pipe into an old pipe. An example of rehabilitation would be relining the pipe. Other examples of rehabilitation include: lining a storage tank, replacing the insides of a fire hydrant, or relining a manhole.

Not every asset can be repaired or rehabilitated. Some assets either cannot easily be repaired or it is not worth the cost to make the repair. An example of an asset that generally does not undergo repair is a residential customer water meter. These assets may not have this particular phase of asset life. They would simply be replaced. Maintenance strategies tie into decisions regarding repair/rehabilitation/replacement decisions. In addition to cases where assets cannot be easily repaired, if at all, such as the one described above, there are other strategies to consider. There may be assets that are too important to fail, so they must be maintained at all costs, there may be assets that are intentionally run to failure with little maintenance and in other cases business processes can be employed to determine cost-effective repair/replacement versus replacement scenarios.

> *Potential Interventions:* Repair of the broken asset using appropriate repair techniques; evaluating repair techniques to determine which are the most effective; conducting a study of repair, rehabilitation and replacement options, including the costs and ability to increase useful life; exploring the actions that trigger maintenance (i.e., time-based or use based) to see if an adjustment is warranted; routine maintenance; corrective maintenance; predictive maintenance; corrective maintenance; and documentation of all repairs and rehabilitation activities in a computerized maintenance management systems (CMMS) or asset management (AM) system.

Replacement: Once the asset has been repaired as many times as possible within the boundaries of financial efficiency or if it is not possible to repair or rehabilitate the asset, it will need to be replaced. At the time of replacement, the new asset is installed, and the existing asset is either abandoned in place or removed for disposal. The new asset can either be the same type of asset or it can be improved in some way. The improvement may be a larger redesign that changes the asset into something else or it might be a different kind of asset, different size, or different configuration.

> *Potential Interventions:* Selection of assets for replacement; determination of whether a different type of asset would be beneficial; capital improvement plan (short or long-term); consideration of system redesign; or documenting replacement of the asset in a CMMS or AM system.

Disposal: The final phase in the asset's life is the disposal phase. At this point the asset is no longer in service and must be disposed of. In some cases, the asset may have some salvage value as scrap or may be recycled, but in most cases, the asset will just need to be taken to an appropriate landfill. There are some abandoned assets that the utility may wish to abandon in place. In these cases, the asset is still physically in its original location but is no longer in service. Some assets can only be abandoned in place, for example, wells. Pipe is sometimes abandoned in place and sometimes removed, depending on the circumstances of construction. If the intent is to use trenchless technologies and the old pipe will not be exposed during construction, it is more likely that an abandoned-in-place strategy will be used.

While this phase of an asset's life may seem unimportant, there are potential issues that can occur when this phase is not properly acknowledged. One northeastern utility discovered that several abandoned sewer pipes were connected to its collection system that were unknown. These pipes were discovered through the process of figuring out why the infiltration/inflow (I/I) rate was so high in the system. Through smoke testing, they found the abandoned lines and discovered that they were still connected to the system. They had never been properly mapped nor were they included in any pipe inventory. Once found, it was discovered that the pipes had cracks and other defects that were allowing large quantities of I/I into the system. Once discovered, the system was able to map these abandoned lines and disconnect them from the system which cut down on the total I/I.

In another example of the importance of proper abandonment, a well was abandoned but never properly closed in a drinking water system. Total coliform, and occasionally *E. coli* started to be found in the system during its monthly total coliform sampling. The source of contamination was eventually traced back to the well and it was discovered that the improperly abandoned well was hydraulically connected to one of the operating wells and was allowing contamination to enter the well. Once discovered, the well was properly abandoned and the situation was resolved.

Many facilities leave abandoned parts lying around in the yard or in the back of a building. This practice is not recommended. It can have a corrosive effect on employee morale, give a bad impression to state or federal regulators, or lead to animal issues (animals taking up residence in the abandoned parts). Unusable vehicles are often left on the property as well. The best approach is to dispose of these parts and vehicles in the accepted manner (some parts or vehicles may have to go through government surplus procedures) and remove them from the property. If anything is actually useable, it should be put into storage in an organized manner and its presence in the spare parts inventory should be noted in case anyone wishes to use it in the future. Alternatively, another part of the government or another community may be able to use some of the parts.

The asset ID number needs to remain with the asset even when it is abandoned, whether it is removed or not, but it is especially important if the asset is to be abandoned in place. If a new asset is installed to replace the old one, a new asset must be assigned, but it is possible to use a variation on the ID number of the old asset, if desired. For example, if the utility is using a smart numbering system and the old asset ID number is MH-ElmST-035, the new asset ID number for the replacement asset could be MH-ElmST-035-R1, with the R1 indicating that it was the first replacement. If the asset is replaced again, it can be MH-ElmST-035-R2. Another alternative would be to put a ".1" at the end of the number signifying replacement 1. In this case the number would be MH-ElmST-035.1. Using a system like this, the numbering system can remain intact while giving a new, unique number to the new asset. Some asset management vendors will suggest just numbering the new asset the same as the old one and removing all the old data. That is not recommended. There is too much valuable history lost when the information is removed from the inventory. Also, if the old information is not removed properly, it will get attached to the new asset which can be very problematic. It may appear that an asset has had numerous repairs when in fact it is a brand-new asset.

> *Potential Interventions:* The most important intervention at this stage is to ensure that the asset is marked as no longer in service, including the date it was decommissioned or removed, within the asset inventory and the costs of disposal or decommissioning are recorded. The date the asset is no longer used can help determine "actual service life" of assets over time which is important information for future planning purposes and a variety of other activities, such as assessing maintenance interventions. The costs of disposal may be considered operating, rather than capital expenses, so understanding these costs can be important. Additionally, the inventory should indicate whether the asset was removed and disposed of or abandoned in place. It is important to know if an asset is still physically there to ensure that it is not accidentally considered a live asset. For example, if a sewer pipe is still in the ground, but no longer in service, it is possible that a household service line could be connected to the old, unused pipe, rather than the new pipe. Furthermore, it is necessary to understand decommissioned, buried assets when assets are marked for any dig operations (i.e., the one-call system). If an asset is decommissioned and is in the way of construction, it is possible to remove all or part of it during the new construction.
>
> Another intervention is examining the history of the asset in the asset inventory to determine how the asset failed. Was it related to mortality, capacity, level of service, or financial inefficiency? Did the asset experience the entire life span you expected; did it last longer

or shorter than the expected useful life? What can be learned from an examination of the asset's failure(s)? Is it information that could improve design, construction, operation, maintenance, or repair? How can this information be used to improve the management of other assets of this type?

Funding: Each phase of the asset's life requires money for all different purposes. Initially, the money is for planning and design. Then the money will be for constructing the asset. Once constructed, all the additional money is spent on interventions directly on the asset, including operations, maintenance, repairs, and rehabilitation. At the point the asset is replaced, the costs are now associated with the new asset, rather than the existing asset. The final cost in the asset's life is disposal or abandonment. Understanding these costs can help determine the asset's total life cycle cost so that the entire life cycle costs can be compared.

When choosing which alternative to complete or which asset to use, it is important to compare based on the overall costs, not just the initial construction costs. Choosing an asset or project with a low initial construction cost, but high ongoing operation and maintenance (O&M) costs can cause the facility hardships later on. Considering all costs, or as many as can reasonably be estimated, can help ensure that the best decision is made to achieve the overall lowest life cycle cost of the asset. In this comparison, the two assets or projects must be given equal length; if one will last longer than the other, a prorated portion of the cost of the new asset must be included. If they have different numbers of required rehabilitations within their life spans or different expected numbers of repairs, different cycles of maintenance, all of these factors should be considered in the comparison.

It may be difficult to fully articulate all of the expected costs, but reasonable estimates should be undertaken so that the costs can be understood. Standard economic analysis using Present Worth or Equivalent Uniform Annual Cost can be used to make the comparisons or a similar approach. Once full costs are compared, an informed decision can be made regarding which asset or project is the most advantageous for the utility/community.

2.0 WHY CARE ABOUT THE ASSET'S FULL LIFE CYCLE?

One of the biggest reasons to care about the full life cycle is the understanding that the most influence over an asset at the least cost occurs in the planning phase. The figure (5.2) from Gibson and Hamilton (1994) shows the relationship between influence and cost during each major phase of the asset's life.

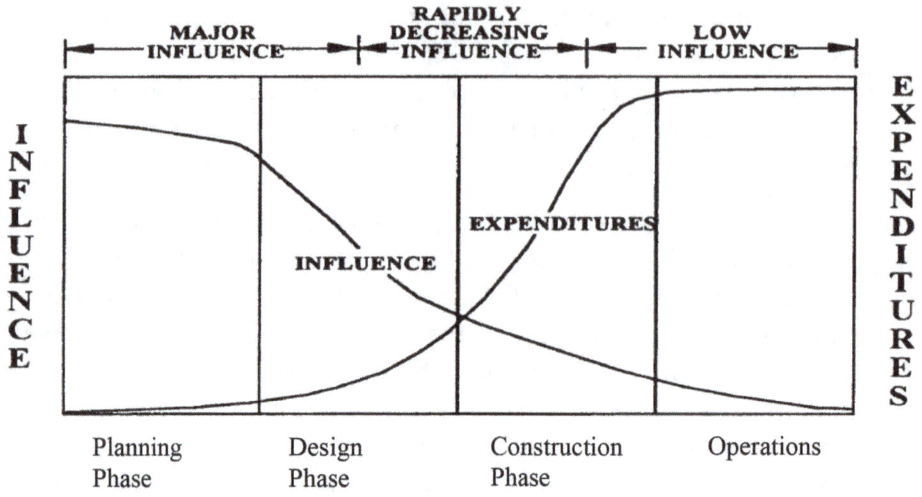

FIGURE 5.2 Cost Influence Curves

Source: Gibson and Hamilton (1994) Analysis of pre-project planning effort and success variables for capital facility projects. Construction Industry Institute Source Document 105. *Note:* The author added the labels "Planning Phase, Design Phase, Construction Phase, and Operations" to the graph. The original does not contain these labels.

During the planning phase, there is tremendous influence over the assets (or project) at a very low cost. This is the time to consider which alternative may be most prudent for the utility and community; what type of assets to use; what material they should be made of; where they should be located; and any other fundamental aspects. At this point, the project is on paper only, it has not been through an expensive design processes, land may not yet be purchased, so making a change at this point costs very little and can result in major improvements over the life of the assets. It can help ensure that the appropriate technology is used, the location is optimal, the materials of construction fit the situation, and the operations personnel will be able to operate the facility.

As the project moves into the design phase, the influence and expenditure lines start to shift. At the beginning of the design process, it is still relatively inexpensive to make changes, but as the design moves from 30% to 60% to 90% complete, it will become more costly to make changes. By the time the design is 100% complete, it will be very difficult to modify the design because one change may result in the need to change many other aspects of the design.

Once the project has progressed to construction, the influence is decreasing substantially, and the costs are rapidly increasing. At this point, just about any change would require a change order which could add considerably to the costs of the overall project. In general, the goal is to minimize change orders to keep costs to the level predicted at the start of construction. However, it is better to address a known problem with a change order than to live with the consequences for 20, 30, 50, or even 100 years of operation. If the problem will cause extra maintenance or decreased asset life, it is most likely worth the cost of the change order. One example is the completion of a wastewater treatment plant in an industrial application in a large, midwestern city. During design, the design engineer miscalculated the height of the outside walkway so that it did not match up with the height of the door on the building leading to the outside. Therefore, there was an approximate one foot drop when a person stepped out of the building. That drop had the potential to lead to injury and should have been addressed at the time of construction. The walkway is likely to last for 30 to 50 years, meaning this defect has been and will continue to be problematic for a considerable period of time.

By the time the asset has reached its operational phase, there is very low influence, and any change would come at a very high cost. At this point, for example, the well is drilled where it is, the storage tank location is set, the treatment type is decided, the operational controls are in place, and changing any of these things could be extraordinarily difficult. Therefore, if one considers the start of the asset's life at the operational phase, there is a lot less that can be done to make the assets more cost-efficient, effective, and sustainable. At this point, adjustments can be made around the margins to fix issues with particular assets, but the overall project characteristics will remain the same until the next major upgrade to the same facilities.

3.0 BLENDING DESIGN AND OPERATION

To address the realities of the expenditure/influence curves along the asset's life cycle, it is important to consider how design and operations can be blended. It is vital that design and operations have a collaborative relationship. Design and operation are inextricably intertwined. The best design in the world will not achieve its intended purpose without appropriate operations and the best operations will not make up for a subpar design. Operations should be considered an integral part of the overall design process. This collaboration should include both internal and external design professionals.

Meaningful integration of design and operations is the intended goal, which can take a bit of a culture change and possibly some training. It may also require upper management to insist that both sides participate. Sometimes, operators are given a chance to input on the 30% and 60% design submittals as the only type of interaction. It can be difficult for individuals with unfamiliarity with design drawings to fully engage and understand the implications of what they are seeing or it may take considerable time to delve into the drawings and fully review them. A more robust process might achieve better results. It is important to consider who is invited to discussions between the utility and the design professionals and at what point in the process these conversations should take place. It is worthwhile to have conversations from the initial scoping of the project through early design to the final submittal. The setting can also have an influence on participation. Holding meetings in locations likely to put the operations staff at ease, such as their treatment facilities or their offices, may improve ability to provide meaningful feedback. The engineers should also seek to provide information in formats most appropriate to operators. For example, perhaps a 3-D design would be better than a flat drawing in cases with vertical (or plant type assets) to help visualize the assets, where they will go, how they all fit together, and how operations staff will interact with the assets. If 3-D drawings are requested at the request for proposal stage of the process, it may not add much to the cost of the overall design.

While, initially, this type of cooperation may feel a bit uncomfortable and may even benefit from a bit of training on both sides or professional facilitation to ease the conversations, it can be well worth the effort. Operators are much more familiar with the facility itself and how the plant operates and may pick up on errors that can be more difficult to detect by someone who has little to no operational experience. As an analogy, consider the process of building a new house. Professionals, either an architect or a builder, are needed to develop the designs and construction drawings for the house, similar to needing an engineer to develop construction drawings for the water assets that are going to be installed. In the case of the house, it is very unlikely that most or all of the decisions would be left to the builder or architect. It is desirable to have input into such decisions as which way the house faces, where doors are located, the size of bedrooms and bathrooms, locations of outlets, and many others. The architect can ensure the electrical wiring will work, but the owner can decide how many and where the outlets are located (within the confines of local codes, of course). It may be worthwhile to think of the water facilities in the same way. The individuals who have to live in the house care how it is designed and constructed. The utility staff has to "live" in the facility for their entire workday, so they care how it is designed and constructed as well.

The relationship being described between engineers and operations staff is a cooperative partnership, not an adversarial one. Rather than "us" and "them" discussions it is a "we" discussion. Both sides have a vested interest in the project succeeding and should act accordingly. The engineer wants the operations to work well and the client to be happy so that they may receive additional work in the future, serve as a positive reference for the engineer, and because they want their design to successfully achieve its intended purpose. The operations staff want the project to work well for obvious reasons—they will have to perform all the interventions in the asset's life once the facility is turned over. They also want their job tasks to be as efficient and effective as possible.

Some key considerations of integrating operations and design include:

- Enabling O&M personnel to discuss their recommendations with designers that facilitate operation and maintenance activities, such as location of equipment, the ease of access to equipment, and potential monitoring and maintenance equipment additions.

A few examples of improper integration of design and operations include a wastewater pre-treatment plant that required an operator to climb inside the containment basin to open and close a valve to add a chemical to the wastewater. Had the valve failed, the operator would have been exposed to a dangerous chemical and would have been standing within the chemical containment basin creating a serious health and safety issue. While the design technically worked, it was not an acceptable operation. The valve location was ultimately moved to solve the issue. In another case, a flow meter at a water plant was placed so high in the building that it could not be read without a ladder. Again, while the fact that a meter was installed technically fit the design criteria, it was not useful from an operations standpoint and the meter was not read routinely because of this placement.

- Discussing the major activities that will be required in the routine operation of each major asset included in the design and the overall operation of the complete facility.
- Determining the expected routine and preventive maintenance activities associated with the major assets. Considering the expected failures with the assets and how long they are anticipated to last in service before expected repair, rehabilitation, or replacement are other important aspects.
- Think through the expected predictive maintenance techniques that can be used on the major assets and their associated costs and necessary equipment and supplies.
- It is critical to know whether the new project will require any change in operator certification levels, numbers of operators, or schedule of when operations staff must be present. (A drinking water facility in the northeast portion of the United States decided on a design that would permit the facility to remain within its existing footprint but

did not thoroughly consider the fact that the new design required personnel to be at the facility 24 hours per day, 7 days per week. The old facility only required operators to be present 8 hours per day with an on-call weekend schedule. Personnel indicated that if they had fully understood the ramifications of the change, they most likely would have selected a different alternative.)

- The spare parts that will be required to address the expected failures, including where they can be obtained and the associated cost. (A wastewater facility in the southwestern United States decided to use ultraviolet [UV] disinfection rather than chlorination/dichlorination. When the light bulbs needed to be replaced, the operators found out that there was only one manufacturer who could supply them for the particular asset they had and that manufacturer was in Europe. Therefore, the replacement bulbs were extremely expensive. They were unaware of this issue prior to selecting the UV system. While having this knowledge may or may not have resulted in a change of vendor for the system, it would at least have allowed the facility to properly budget for spare parts.)
- Along with concerns related to operator certification, it is necessary to understand if the project will require any specialized training for operators or maintenance personnel in order to ensure a smooth start-up and long-term operation. Other considerations include: whether the training can be obtained before the assets are installed, whether there is an additional cost for training, and whether there are any existing facilities nearby that can be visited to learn more about the facility.

The most important question to ask is whether these key operation questions enter into the normal discussions that take place when new assets are being considered, projects are being planned, or designs are being completed. If not, why not? It is far better to understand these considerations *before* assets are in the ground rather than after. Remember, the influence occurs at the beginning of the asset's life during planning when changes can be made at very low cost.

Similarly, it is equally important that the alternatives evaluation process be conducted as a partnership between the design personnel and the facility personnel, including operations and maintenance. There is always more than one alternative to just about anything that will be constructed in the water arena. There may be many alternatives or just a few; in either case, consideration must be given to the alternatives evaluation and selection to make sure that the most appropriate option is chosen. This process may not seem like it relates to asset management, but it most definitely does. The preferred option will determine what assets are installed, which will set in motion all the future operations and maintenance-related activities.

Some questions to ask during the alternatives evaluation process include the following:

- Have all viable alternatives been considered? Are they all given equal consideration or is the preferred alternative pre-ordained? What does the write-up of the alternatives look like? Are they written in a way that it is obvious which is preferred? What does the costing portion of the analysis look like? Were all viable alternatives given the same level of costing?

A utility in the southwestern U.S. had the selected alternative chosen exclusively by the engineer and it included a reverse osmosis (RO) system. This utility did not need an RO system, as there were no specific contaminants that needed to be removed. The RO unit was expensive to operate, energy intensive, and caused numerous other issues, such as pitted copper piping and lead and copper regulatory violations. It was very difficult to remove the system once installed, and the operator did not want to have to tell his bosses that a bad decision had been made during the planning phase of the project. It would have been better to thoroughly consider other alternatives.

- Can operations personnel suggest alternatives to consider? If so, are these options given equal weight during the review process?
- Are full life cycle costs of the alternatives taken into consideration or just the initial costs?
- Is energy use/energy efficiency/energy source taken into consideration when the alternatives are compared?
- Are new technologies being considered or just standard technologies? If new technologies are being considered, are there sufficient examples of the use of the technology in existing facilities to provide a comfort level that it will work? Is it possible to speak to any of the facilities using the technology?
- How is the ultimate decision on the preferred alternative going to be made? Who has input into that decision? Can the operators, managers, or other staff provide feedback to the decision-makers prior to the decision being made? If so, in what way?

It is important to think about past projects and how successful this part of the process has been and whether there has been overall satisfaction. It is also necessary to consider whether the upper management or elected leaders would have a different answer than the operations staff. If past projects have not always received the best advice or installed the best alternatives, it is worth considering a new working relationship to improve the process in the future. As discussed at the very beginning, it is not possible to change the past. The past can be used as a learning experience for how things can be different in the future. If project selection has not been as favorable to the utility as hoped, make changes to improve it in the future.

A Case Study for Interaction Between Engineers and Operations

A large utility in the midwestern U.S. developed a very positive relationship between its operations staff and its engineers. As part of this process, a 3-D design was created to allow the operators to virtually operate the facility. This process allowed the operators to identify some errors in the design that could be addressed before the project proceeded to construction. Additionally, the operators were able to make suggestions regarding improvements to the overall design that would facilitate easier operations practices. These changes were made to the design before it was built. Additionally, the operators were able to visit treatment plants that had the same type of technology and receive training on operation and maintenance before start-up. By the time the facility was put in operation, the operators were very comfortable with the equipment. There were fewer change orders than with past projects as well as fewer maintenance issues after start-up. The entire process was smoother and more cooperative. Everyone involved was pleased with the outcome. An operator at the facility noted that most of the time staff is handed a facility and told to make it work, but in this case, they were involved the entire time and it was their facility. The design portion of the project was a bit more expensive than their traditional process, given the 3-D design and some extra meetings, but the overall costs were much lower given the other cost savings from fewer change orders, better asset operation, and fewer start-up problems. There were also intangible benefits from the high level of operator satisfaction based on their involvement in planning and design and the higher comfort level they experienced with the new operation. This utility intends to use this type of approach for all future designs.

4.0 ASSET MANAGEMENT VS. MANAGING ASSETS

Life cycle costing is the intersection of asset management and managing assets. Asset management is thinking about the collective and what is strategic for all assets taken together or at least what is strategic for collections of assets. Managing assets is thinking about individual assets and what is advantageous in the management of each of them separately.

TABLE 5.1 Asset Management or Managing Assets

Activity	Asset Management or Managing Assets
Development of an Operation and Maintenance Plan	Asset Management
Preventative work order to exercise a valve	Managing Assets
Work order system to track proactive vs. corrective maintenance	Asset Management
Timeframe to lubricate all pumps in the system based on a study of lubricant and pump performance	Asset Management
Repair on a piece of pipe	Managing Assets
A study of break repairs on all pipes to assess the effectiveness in extending pipe life	Asset Management
Replacement of an inefficient pump with a more efficient pump	Managing Assets

Both functions are important, but they are not the same thing. With asset management, the organization considers its overall strategic direction, the goals it is trying to achieve, its overall business risk, and how it intends to manage it. Asset management considers the big picture of what the organization is trying to accomplish and how the assets will be strategically managed to achieve the intended purpose.

With managing assets, the organization is considering how best to operate, maintain, and intervene during the life cycle of an individual asset. It is important to recognize the difference because focusing only on individual assets will not achieve all the benefits of an asset management program. At the beginning of asset management, managing assets is often the focus, but if the importance of asset management can be understood, it is possible to move beyond the individual assets to the collective and achieve far greater benefits. Some examples of asset management and managing assets are shown below in Table 5.1.

In looking at the items on the table, consider why each is in that category. Think about how the "asset management" activities are strategic and about the collective, and how the "managing assets" are about the individual assets.

5.0 OPERATIONS

It is important to remember that operations and maintenance—the two parts of "O&M"—are two different, but interrelated activities. Operations entail the day-to-day activities to enable the assets to do the jobs for which they were intended for the maximum amount of time. The goal is to maximize "uptime" which is the time the asset is operating in the manner intended or is available to operate in cases in which operation is intermittent. The more the asset is doing its job, the better off the facility and the more likely that the associated level of service goals can be met and compliance can be achieved.

To maximize uptime, it is necessary to consider the effect of operations, and whether any improvements can be made to maintain or increase uptime. A few examples of the relationship between operations and uptime are presented below, but there are many others.

- For pumps, avoid frequent starts and stops, if at all possible, as starting and stopping are times that put a lot of stress on pumps. If there are frequent starts and stops, the pumps may degrade faster than if there are fewer starts and stops.
- When pumps are on lead/lag operation, the way lead/lag is handled needs to be considered. Lead/lag can be switched each time a pump starts so that the lead pump changes with every start or the lead pump can change on some routine basis where it is switched based on run time or something similar. Alternatively, one pump can remain as lead for a long period of time, potentially until it starts to show signs of distress. The methodology of how lead/lag is practiced changes wear and tear on the various pumps from uniform wear and tear to fairly significant differences in how the pump condition changes. This choice may mean that all pumps fail within a relatively similar time frame or that one pump fails much sooner than the others. Either operation can be successful, it just needs to be considered so the facility can be prepared to act appropriately when a pump experiences a malfunction or failure.
- Another aspect involving pumps is whether to have variable frequency drives (VFDs) or not. Using VFDs can effect a pump's overall wear and tear, condition, and expected useful life by allowing a pump to operate at lower speed when demand is lower. VFDs can also help with energy efficiency. However, not all situations are well-suited to VFDs and they can add cost to the installation.
- For fire hydrants, it is important to shut the hydrant slowly to avoid water hammer. The water hammer can cause damage or stress to other assets, such as pipes and isolation valves.
- The pipe distribution network should be operated to reduce pressure surges and water hammer as much as possible.

- For wet wells in wastewater systems or storage tanks in drinking water systems that are used to start and stop pumps based on level controls, consider the levels that start and stop the pump. Is enough of the capacity being used to start and stop the pumps? Could the levels be adjusted to have an improved operation and better use of tanks and pumps? In the case of storage tanks, it is important to consider the amount of water turnover within the tank to keep water age within an acceptable range. With wet wells it is important to consider how fast the wet well rises during storms or significant I/I events and ensure that the pump can keep up with the incoming flow. If not, the wet well level may need to be lower to accommodate a bit more free board to prevent overflow conditions when flows are high.

If we want to maximize uptime, it is important to consider what might interrupt uptime. There are three main items that interrupt uptime: unexpected failures, expected failures, and planned maintenance (Figure 5.3). Each of these disrupters are described below.

Planned Maintenance: An asset is likely to need some kind of planned maintenance at one point or another during its life. This planned maintenance may be something like lubrication, cleaning, testing, or inspecting parts, and as its name implies, it includes only activities that are *planned*, not activities that are unexpected, unplanned, or emergency in nature. Because these activities are planned, there is a large opportunity to reduce disruptions to uptime. Some examples include:

- Making sure spare parts are available to finish the job quickly the first time without having to leave to get more parts or stop partway through the job.
- Ensuring that any other necessary supplies are available, such as lubricants and personal protective equipment.

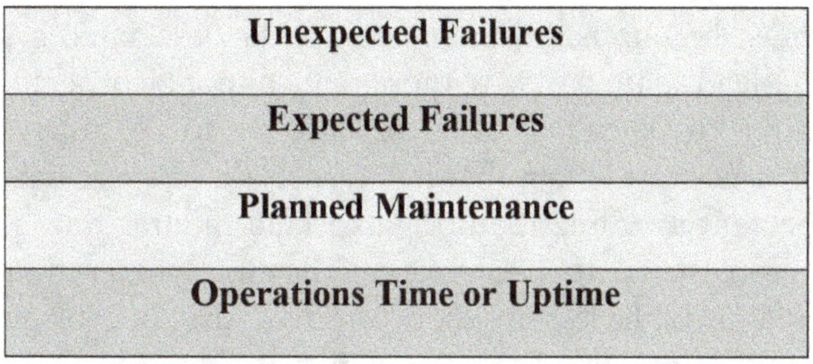

FIGURE 5.3 Uptime and Its Disrupters

- Making sure that necessary staff are available and ready to do the job before beginning.
- Making sure that any safety procedures are undertaken, such as lockout, tagout.
- Timing the maintenance task to cause the least amount of disruption.
- Estimating the time to complete the job and ensuring that you have the time to complete it before the asset is needed or before shifts end. It is important that the job be scheduled for the amount of time necessary to do the complete job.
- Investigating past activities to determine the optimal timing and approach to the maintenance task and looking for any issues that have occurred in the past so that they can be addressed before starting this maintenance task.

It is also important to question assumptions related to how and when to do planned maintenance. It is important to remember that just because "we've always done it this way" does not mean it has to be done this way. If there is a better way, it should be explored.

Expected Failures: With just about every asset, there are ways that the asset may be expected to fail. These are things that are known to wear out and cause issues. These failures include such items as a bearing, an O ring, a joint, or mortar within a larger asset in which these items may fail while the overall asset may be okay. These items will need to be replaced as they fail to keep the asset in service.

It is important to be prepared to address these failures with spare parts, needed equipment and supplies, necessary staff, and proper training so that when the expected failure occurs, it can be addressed quickly. For highly critical assets, it is extremely important to consider the expected failures and to be prepared to address them immediately upon failure. Long-lead spare parts related to expected failures may need to be purchased far in advance of the failure to ensure that the extensive order time does not delay the repair and must be purchased in advance if the associated asset is highly critical. (Long-lead spare parts are those that cannot be ordered, purchased, or delivered quickly. They may take weeks or months to receive because of supply chain issues or the fact that it is a specialized spare part.)

It is important to think about the expected failures that an asset may experience before their actual occurrence. The more preparation is made for the expected failures, the less uptime is affected. Also, in considering known failures, it may be possible to design some of them out. As an example, if failures occur because a particular type of material was used, another material can be used instead to prevent the failure. For example, a system whose assets were routinely subjected to a harsh environment, found that they had many failures in metal components. When the metal components were replaced with stainless steel, many fewer failures occurred.

Unexpected Failures: While expected failures are anticipated, unexpected failures, by definition, are not, so they are hard to prepare for ahead of time. For unexpected failures, it is important to be prepared to react as quickly as possible. Steps such as knowing how to contact the manufacturer and supplier and where and how to get spare parts for that particular asset can help speed up recovery from unexpected failures. For highly critical assets, it is also possible to have a full spare waiting to be installed so that if an unexpected failure happens, it is possible to install the spare immediately and return the asset to service. Another approach is to ensure full redundancy is available so that another asset can do the job until this asset is repaired.

If the same type of unexpected failures occur multiple times, it is no longer an unexpected failure and moves to an expected failure. At that point, it is possible to prepare for this type of failure in the same way as described in the expected failure section. Consider the case of a large drinking water system that was experiencing premature failures on its polyvinyl chloride (PVC) pipe. The pipe was failing much sooner than anticipated and for most systems, an early failure of this type would be completely unexpected, so this failure incident would fall under "unexpected failures" for most. However, given that the failures were repeatedly occurring in many of the relatively new PVC pipes, they are no longer "unexpected" for this system, but rather "expected." The failures were thought to be related to design and construction issues, so the system is trying to design out the failures in the future by changing the installation practices. Alternatively, they could choose a more durable pipe type that can handle their harsh underground environment.

If lots of unexpected failures happen (in this case these are different types of failures on the same asset, not the same type multiple times), it may be necessary to consider designing the asset out or choosing a different type of asset. Multiple unexpected failures can be too disruptive to the operation to allow this to continue, especially if the overall criticality of the asset is high or moderate from a consequence perspective.

In all cases, it is necessary to collect good information through a robust work order system that can track occurrences on each asset. Without this information, it is difficult to know what types of events have disrupted uptime, how uptime has been disrupted, and what actions have been and can be taken in the future. The goal is to ensure uptime represents the majority of time for the asset and the other three categories are much less. Ideally, within the disrupters, the planned maintenance would be the largest category and unexpected failures would be the smallest category, hopefully very small. Figure 5.4 shows the desired comparison.

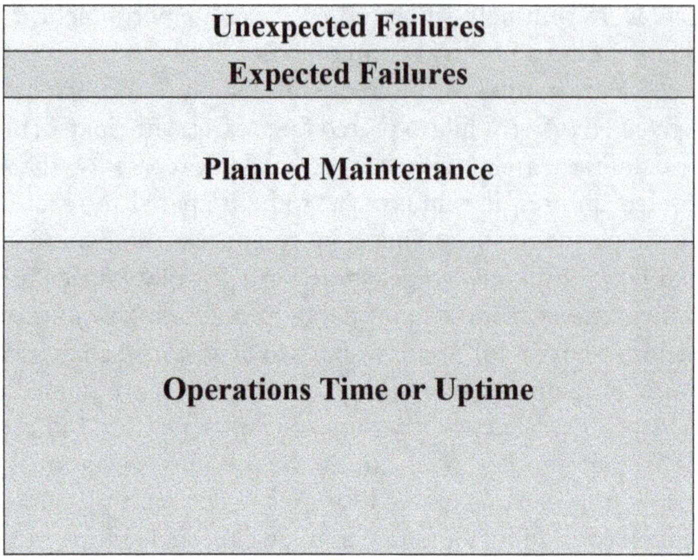

FIGURE 5.4 Desired Uptime and Its Disrupters

Overall Operation: Beyond the operation of individual assets, how the system is operated as a whole can affect service to customers and asset life. Some examples to consider are presented below.

- Automatic switchover to generators to prevent shutdown over lack of electricity
- Changes to redundancy to create more redundancy and reduce potential for down time in highly critical assets
- Operational training so operators have the knowledge they need to efficiently and effectively do their jobs and perform their assigned activities
- Ensure operator vacancies are filled as soon as possible with qualified personnel
- Understand the flow path of the water or wastewater system
- Provide written instructions for emergency operation
- Ensure safety training is provided to all employees and that all personal protective equipment is available, accessible, properly fitted, in good working order, and stored properly.

The goal with operations is to improve the reliability of the system and reduce down time so that the levels of service are met as often as possible, including regulatory compliance.

6.0 MAINTENANCE

The previous section talked about the "O" in O&M, this section will discuss the "M" or maintenance. There are a few different types of maintenance that can be performed. In general, maintenance includes routine maintenance, preventative maintenance, corrective maintenance, and predictive maintenance. However, different people can use slightly different terminology so other terms may be used to mean the same things as those being discussed here. Also, routine maintenance can be combined with preventative maintenance in the same category.

Routine maintenance includes those activities that are performed to keep the assets in proper operating condition, such as cleaning, inspecting, and lubricating. Preventative maintenance includes tasks that are done to prevent an asset from reaching potential or functional failure. These activities can include valve exercising, adjusting bearings, realignment, sewer cleaning, and pigging a water line. Corrective maintenance includes actions taken to restore an asset after it has failed in some way. It can include any kind of repairs made to the asset. Predictive maintenance includes tasks that attempt to determine when an asset might fail. These activities can include: thermal analysis, vibration analysis, oil analysis, leak detection, non-destructive pipe inspection, closed circuit televising (CCTV) of sewer lines, storage tank inspections, and electrical insulation testing.

Conducting maintenance of any type takes resources—equipment, supplies, and personnel—which all cost money. However, it is wrong to think of maintenance as a cost rather than an investment. The investment in maintenance ensures that assets will achieve their lowest life cycle costs. Without it, every asset will run to failure, including high consequence assets, which will cause many negative effects to the organization, including poor customer service, demoralized workforce, constant disruptions to uptime, and increases in safety incidents. While it might seem like taking maintenance completely out of the budget would make it less costly for the organization to run the utility, that is not the case. Operating in a completely reactive way, without doing proactive work, is much more expensive. Waiting for an asset to fail can lead to that asset causing cascading failures and more parts failing at once instead of a simple repair when a problem is first observed. For staff, running from one failed asset to the next without ever getting ahead of the issue can be very unsatisfying. Workers may have to move past an asset that is showing signs of distress to get to one that has already failed. It may lead to poor operator retention, which is not positive for anyone.

One goal with asset management is to improve the ratio of proactive and reactive work. Every utility has a mix of both. No matter how much maintenance is performed, some assets will fail and, as discussed in the previous section, some assets will experience unexpected failures. Additionally, some assets are intentionally run to failure when this is the best asset life cycle approach.

A large utility, before embarking on an initiative to improve asset management practice, was acting in a completely reactive mode. Just about every asset was run to failure. Preventative maintenance could not be done because there was no system in place to set up work orders, there was no tracking of what had already been done, and there were too many failed assets that demanded immediate replacement. Employees were walking past assets with potential failures to get to assets that had already failed. In one case, a mixer was making noise due to impending bearing failure and was leaking. The asset was ignored due to higher priorities. The bearings went bad due to a lack of oil in the asset. The oil would have been a very inexpensive fix, but taking the asset apart and replacing the bearings was much more expensive and time consuming. The mixer was out of service for an extended period during the failure and subsequent fix. The utility determined that no oil had been added for more than a year. The system has since changed their practices and now has a thorough preventative maintenance program in place and the balance of proactive to reactive work is much closer to the goal of 75/25.

Routine and preventative maintenance are less expensive than corrective maintenance and emergency repairs. It is cheaper to intervene before an asset breaks rather than after, which gets back to the idea of maintenance as an investment, not a cost center. Proper maintenance can increase uptime, decrease failures, extend asset life, and improve overall operability of the facility. While it may not be possible to do all the suggested routine and preventative maintenance on every asset, it should be possible to address the most important assets and the most important tasks, and budget and personnel adjustments should be made if it is not possible to accomplish these tasks.

Predictive maintenance can require specialized equipment which may be costly, specialized training to understand how the equipment works and how to use it properly, or potentially an outside contractor to perform the work. Because this work can be expensive, it is not generally performed on low-risk assets. This activity is usually reserved for higher-risk assets to better plan for rehabilitation and replacements and for moderate risk assets that have a high consequence of failure. Understanding deterioration can allow a utility to replace an asset before it fails, when that asset failure would be catastrophic. Predictive maintenance can also be performed to reduce the amount of replacement work. For example, if a pipe is slated for replacement, a condition assessment can be done of the pipe to determine the interior condition. If the interior condition shows some sections with failures or weak points and others that are in good shape, less of the pipe can be replaced than otherwise would have.

Routine condition inspection of sewer pipes is fairly common and is generally done on just about every sewer pipe in the system. However, the schedule may vary based on criticality and other factors. For example, a pipe that is highly critical may be inspected every year or every few years whereas a pipe that is in better condition may be inspected every 10 years or so. For water pipes it is much less common to conduct an inspection of all pipes in the distribution system. While sewer inspections are quite likely to be done by the utility itself, especially for larger ones, water pipe condition inspection usually requires an outside contractor. The cost of doing an entire system is generally too high to be worthwhile. Also, many of the pipes in a water distribution network are serving residential areas and are not typically high consequence of failure assets. For larger transmission mains; pipes under a major roadway, waterway, or railroad track; a pipe from the water source to the system; pipes serving critical customers; or a high pressure pipe whose failure may cause considerable damage, the utility would likely benefit from information regarding the pipe's condition. This data may be collected once or a routine investigation may be done every so many years depending on the results of the condition assessment. It is important to specify in the contract or agreement with the outside vendor or contractor the desired data format and require the data to be delivered in this manner before final payment. As with most things in asset management, it is about balancing the available resources so that they can be deployed in the best way possible to have the biggest effect.

Because it is not possible to do all maintenance that could be done, it is necessary to use other aspects of asset management to help determine what to do. One aspect is how maintenance relates to criticality and an example of balancing the costs and benefits of the various types of maintenance is shown on Figure 5.5.

For high-risk assets (the box on the upper right), using predictive maintenance can help determine the right time to replace an asset, how much of that asset to replace, or how best to replace or rehabilitate the asset. Once these assets are slated for replacement, it is not worth doing a lot of extra interventions because the asset is being removed and all of that work will be wasted. However, the asset must have sufficient interventions to keep it from failing before its replacement date.

For moderate risk assets that are unlikely to fail but have a high consequence if they do it (box on the upper left), predictive maintenance can be worthwhile to determine when the asset is starting to deteriorate. Other interventions, such as preventative and routine maintenance should

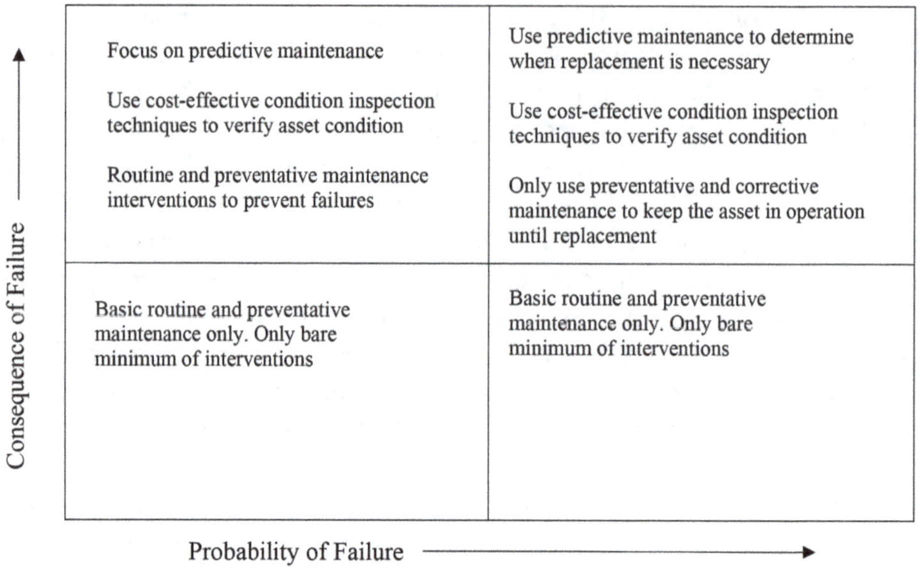

FIGURE 5.5 Repair/Replace/Rehabilitate and Criticality

be done to keep the asset from moving into the high-risk box and to prevent an asset failure.

It is worth remembering that we do not want high consequence assets to fail, so interventions that prevent the failure are worth the expense and resources.

On the other hand, those assets in the bottom two boxes have low consequences of failure. For those assets that are in the lowest risk box (bottom left), only a bare minimum of interventions are required. We want to do what is necessary to keep the asset running and maintain its useful lifespan, such as lubrication, cleaning, inspections, but we do not need to spend a lot of money on other more expensive interventions such as predictive maintenance. We can run these assets to failure in a managed failure mode. The assets in the moderate risk box on the bottom right can also be run to failure and those interventions that extend the life as long as possible without expending considerable resources should be undertaken. The overall amount of maintenance to be performed is a balance between asset criticality, cost, and resources.

Another part of maintenance is to consider when to intervene during an asset's life. When the asset is installed, it will deteriorate over time along a curve. The curved line in Figure 5.6 below represents the deterioration. While it is shown as a smooth curve, it is unlikely to really decay in such a smooth pattern. The curve is being used for illustration purposes. At the start of the asset's life its condition is the best it will be.

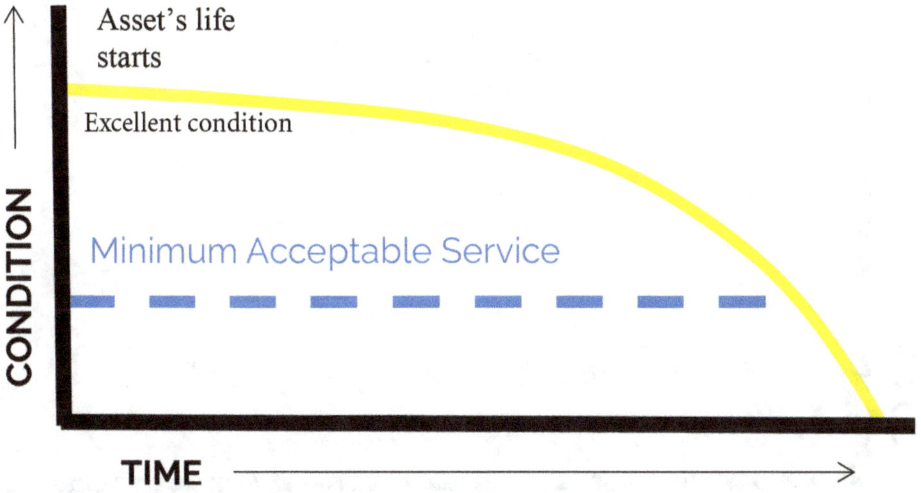

FIGURE 5.6 Asset's Life Begins

As time goes on, the asset deteriorates from excellent or great condition to good condition (Figure 5.7).

As the asset continues to deteriorate, it will reach average condition and it may begin to show signs of failure. These signs might be noise or vibration or some other outward sign that something has gone wrong with the asset. When these signs occur, this is point "P" or a Potential Failure, on the figure below (Figure 5.8).

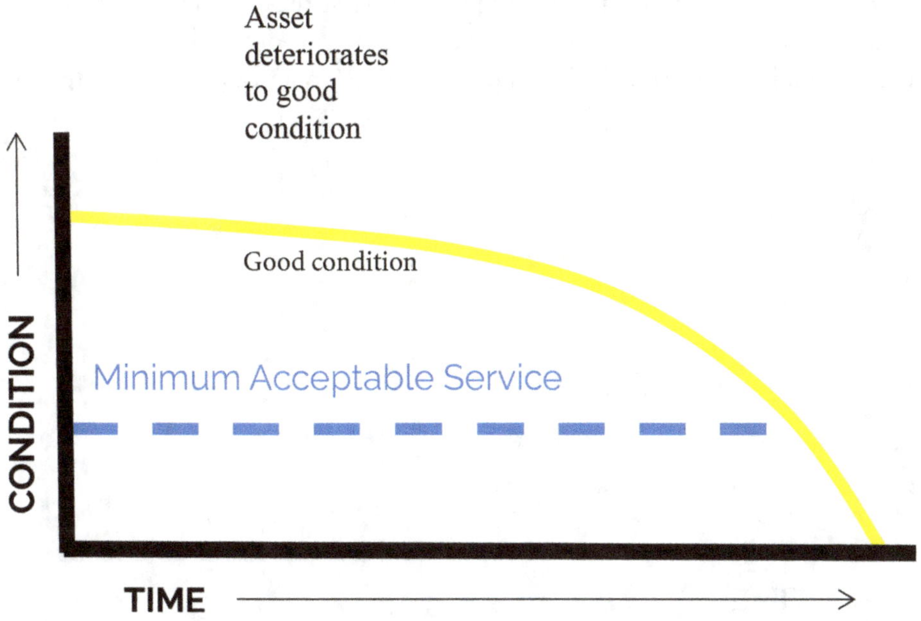

FIGURE 5.7 Asset Deteriorates Over Time

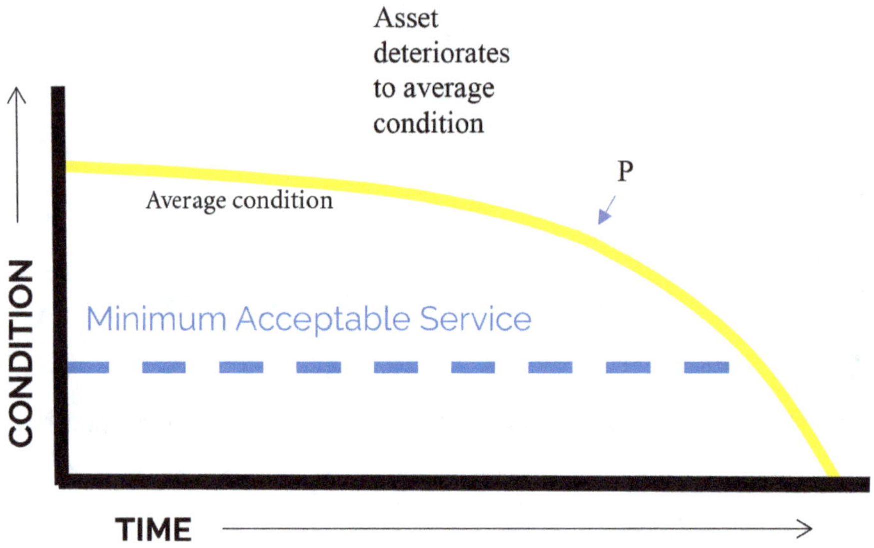

FIGURE 5.8 Asset Deteriorates to Potential Failure

If we take action before the potential failure occurs, it may be possible to prevent it from happening. In this case, the action is not being taken because the asset is showing signs of wear and tear but rather based on some other conditions, such as time based (every X months or every Y years) or activity based (every X hours of service or Y gallons processed). If instead the asset deteriorates to the point of the potential failure, action can be taken to restore the asset to a good condition and prevent the asset from continuing to deteriorate or the asset can be allowed to continue down the path towards a functional failure (Point F on Figure 5.9). A functional failure happens

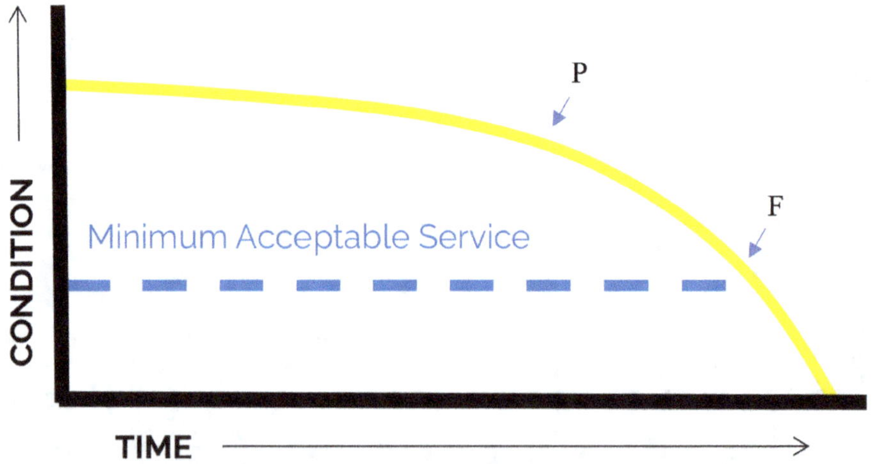

FIGURE 5.9 Asset Deteriorates to Functional Failure

when the asset is no longer performing its intended purpose. The asset may be doing something, but it is not doing that function at the level desired. For example, a sewer pipe may experience a major hole that allows much of the wastewater to flow out of the pipe instead of enabling it to continue down the sewer pipe to the pump station or treatment plant. The sewer is still functioning, but it is not functioning at the desired level; i.e., having raw wastewater leaving the pipe is not acceptable and must be addressed. Some assets, upon reaching functional failure, will not operate at all.

If we don't intervene at the point of the functional failure and restore the asset to a better condition, the asset will continue to deteriorate, probably quickly, to a point where there is a catastrophic failure (Point C on Figure 5.10) and the only option will be to replace the asset. It is possible that at the functional failure stage, replacement is the only option as well, but there may be other options. Going back to the sewer example, if the pipe is not fixed, the sewage leading out of the hole may cause a sinkhole and an entire segment of pipe may fail and other assets, such as streets and sidewalks, may also fail. In either case, at the point of functional failure and below to catastrophic failure, the asset is no longer providing the desired service level and overall functionality will suffer.

The costs rise the longer the personnel take to intervene. The sooner action is taken, the lower the costs overall (see Figure 5.11).

Every action taken before the potential failure is considered proactive work (see Figure 5.12). In these cases, the interventions occur before something happens. These interventions will include routine maintenance and preventative maintenance performed on a time or action basis. The action

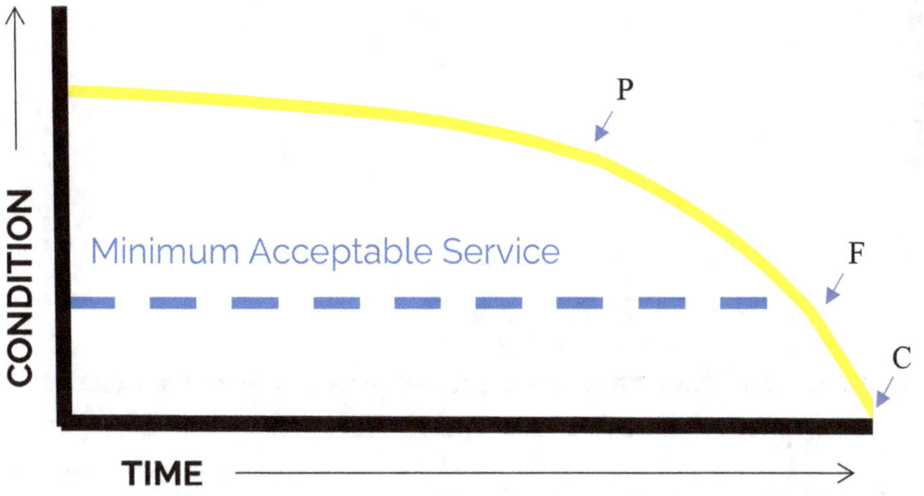

FIGURE 5.10 Asset Deteriorates to Catastrophic Failure

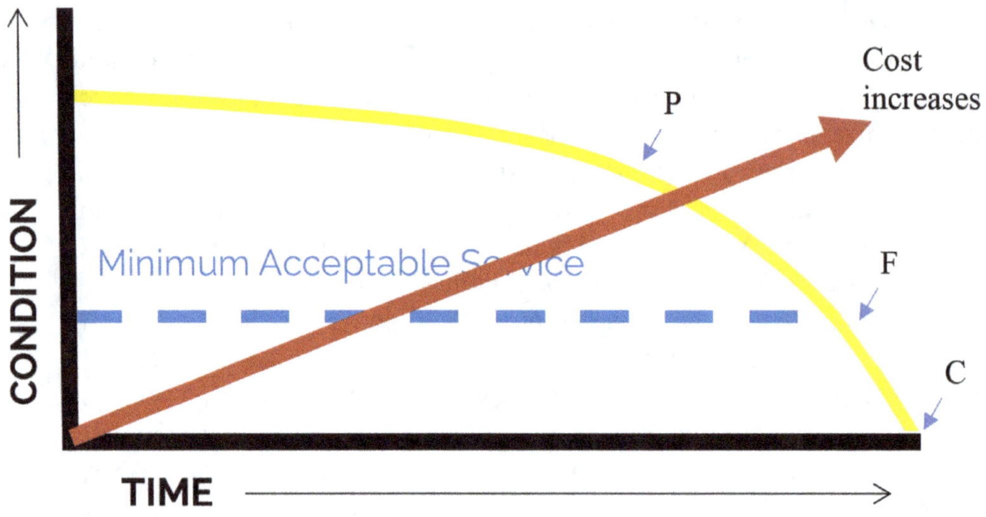

FIGURE 5.11 Costs of Intervention

is taken, not because an outward sign has indicated the need for action, but because time or use have initiated the intervention.

Any action taken after the point of potential failure is considered reactive work (see Figure 5.13). These actions are corrective maintenance actions taken based on an outward sign of some type indicating there is an issue or failure. To maximize uptime in operations and make sure assets are

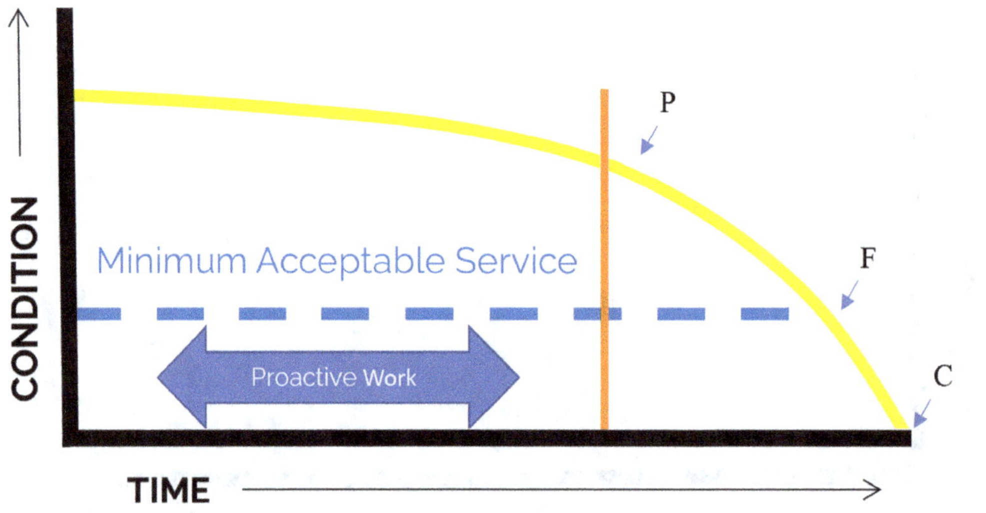

FIGURE 5.12 Proactive Maintenance

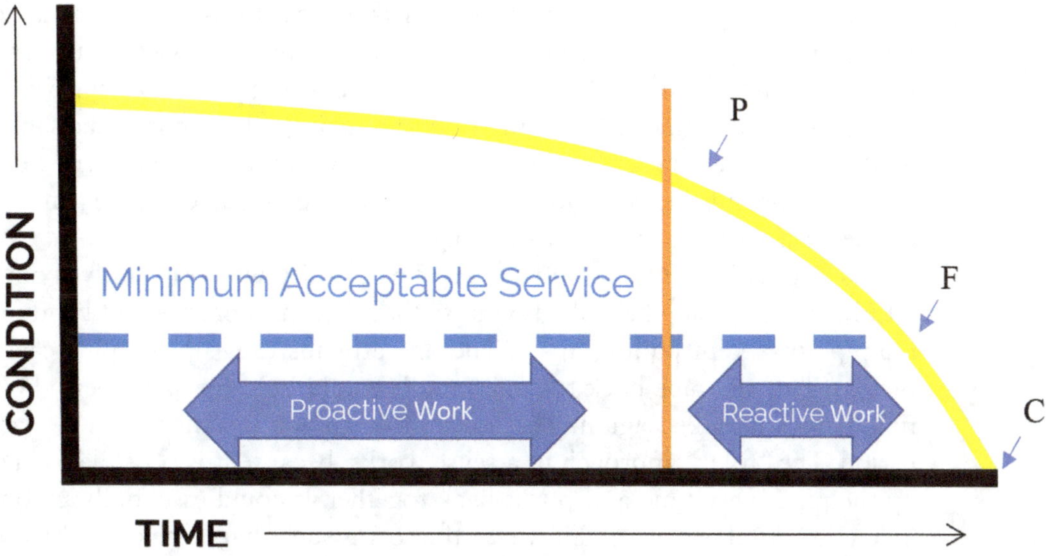

FIGURE 5.13 Reactive Maintenance

operating cost-effectively and efficiently, it is desirable to have more proactive work than reactive work. A goal is somewhere between 70% and 80% proactive work to 20% to 30% reactive work. If the ratio becomes much more on the reactive side, something like 10% to 30% proactive to 90% to 70% reactive, the staff will constantly be addressing failed assets and running from one issue to the next. This situation prevents personnel from being able to do proactive work because it is not possible to do that work if there is a failed asset that needs immediate attention. It is costly in terms of dollars to operate in this mode as well as negatively affecting staff. It is unsatisfying for the employees and makes for an unplanned day when the activities are mostly reactive. It takes an investment in money and personnel to get over the hump and start shifting the balance back to more proactive and less reactive work. After the shift occurs, the employees should see a much more planned workday and the overall cost should go down.

The ratio of proactive to reactive work refers to the amount of maintenance tasks completed, not to the cost. Proactive work will be less expensive for the most part than reactive work so the ratio of expenditures may be different than the ratio related to the number of work orders in each category.

Another aspect of maintenance that is an important consideration is the concept of "don't just fix it, improve it" which was developed by Winston P. and Winston J. Ledet and Sherri Abshire (Ledet et al., 2012), In this concept, it is important to think about whether the actions taken will make the situation better or whether they will just restore the asset to a point where the same failure will occur in the future. If the asset is "improved" on the other hand, the failure will not occur again (at least not in the same way or the same type of failure). Because we want assets to perform well and meet customer expectations, the idea of improving assets is very much in line with asset management thinking.

As one real-world example, a very small utility was constantly fixing a leaking pipe. The pipe leaked because there was an unrestrained bend in the pipe that kept pulling apart. The operator just dug up the pipe each time, pushed the pipe back together and it would hold for some period of time before it leaked again. There were five known instances in a 2-year period. The "fix it" approach in a sense works, because it is possible to dig up the pipe every time and put it back together. It could essentially go on indefinitely in this manner. However, the cost is such that it makes no sense to continue down this path when a better option is to properly restrain the joint so that it does not pull apart in the future. The "improve it" option allows the operator to spend their time on another activity rather than routinely addressing this issue. Imagine if the joint had been restrained the first time the leak occurred, rather than just "fixed," the other four breaks in the 5-year period would not have occurred. The cost avoidance (the cost of the four breaks that did not have to occur) for this particular utility was $5,600. While this figure may not seem like much, consider that the utility served a few hundred homes and had a very small budget and a part-time operator. The cost avoidance was quite significant for them. Furthermore, if the situation was not addressed, it would have continued year after year, adding up to a considerable sum.

In another real-world example, a utility was located in a harsh environment. Regular metallic bolts were routinely corroding and causing problems for the utility. Switching to stainless steel bolts allowed the utility to more easily address future maintenance and the bolts did not have to be replaced on a routine basis.

For any failure that is repetitive in nature, there may be an opportunity to consider whether there is a "improve it pathway" instead of just a "fix it pathway." To find these opportunities and act on them, it is important to have an open line of communication between managers and field staff. The field staff can suggest options and opportunities to make these kinds of changes.

Opportunities With Maintenance: Utilities often do things a certain way because they have always done it that way. Activities were handed down from operator to operator without a lot of thought about why things were done that way or why they were done at all. Asset management is a chance to take another look at activities—what is done, how it is done, why it is done, when it is done—and consider other alternatives.

One large utility was replacing the lubricant in their large pumps every year. They sought a better lubricant—a food-grade lubricant—and found one that only needed to be changed every 4 years instead of every year. This change resulted in triple bottom line (TBL) benefits associated with the lubrication change. Before the change, 54 gallons of lubricant was used for each treatment train change, 16 staff hours were used, and the total annual cost was $2,100.

The practice started in 2001 as each treatment train was taken offline to do preventive maintenance and it continued annually up until 2014, which would amount to a total of $29,400. Starting in 2015, the practice was changed to be once every 4 years instead. The cost from 2014 to 2024 was $5,200 for a total savings of approximately $24,200. In addition to the cost savings, there was also an environmental benefit in that zero gallons are used each year for 3 years and lubricant is only used every 4 years. Over a 4-year period this is a savings of 162 gallons of lubricant. Also, there was no disposal of spent lubricant for 3 out of 4 years. Finally, there were social benefits in that a difficult job did not have to be done 3 of 4 years. Only two staff hours were needed in the 3 of 4 years that lubrication changes were not conducted, with the time relating to pulling samples to test them to make sure the lubricant was in good condition and did not need changing. Over 4 years, this was a savings of 42 hours, or slightly more than 1 week. The benefits may seem relatively small for a big utility, but every hour not spent doing a job that does not need to be done is an hour that can be spent on a more worthwhile activity. Also, if several of these approaches are taken in the same utility, significant cost and resource savings can ensue.

In another example, the staff of a utility had to make several trips a week to get small parts to complete repairs. These trips would occur multiple times per week and take approximately 1 hour per trip. This could amount to 52 to 104 hours per year of wasted time. Instead, the utility hired a vendor to create a vendor-managed inventory of small parts. The parts are always available and the vendor ensures that each part is filled within the level of required minimum and maximum. This keeps the total time of a repair to a minimum and saves money overall by reducing unnecessary trips and hours.

If any monetary or time benefits are realized from looking for and undertaking these opportunities, it is time to think about what can be done with the savings. There are many options from additional AM activities/improvements to the overall AM program, more routine or preventative maintenance activities, completing some projects that are below the funding line or projects important to employee satisfaction that are not on list (e.g., automated meter reading), replacing an individual asset or several assets, or using the time or money for training or other professional development.

Some changes are not as much cost savings as time savings. For example, if there is a map to locate assets such that it is not necessary to spend a lot of time looking for assets, the extra time can be spent on something else. In addition to maps for horizontal assets, field guides can be tied to the asset inventory for vertical plant assets to help operators and maintenance personnel locate these assets quickly during necessary operation and maintenance activities. These can be particularly helpful in large, complex plants or in cases in which small spaces make it more difficult to follow or locate assets. Similarly to the idea of what to do with the extra money, it is important to consider what to accomplish with the extra time. Saving money or time in one area is about using it in another area. It is not about reducing the budget overall or reducing staffing, because every utility faces a backlog of work that needs to be done and not enough time or money to do it all.

Reliability and Maintenance: Reliability is having the maintenance program ensure assets and systems will perform as intended continuously over the long term (i.e., reliably). It is a measure of how well the maintenance program maximizes uptime and how well maintenance keeps assets running. Reliability centered maintenance (RCM) is a preventative maintenance approach to optimize reliability. It involves determining failure modes and ways to prevent them before they happen. If failure modes are known, actions can be taken to address them before they occur.

There are many metrics available to measure maintenance and reliability. Some of them are presented here.

- MTTR is the Mean Time to Repair. This metric is the time it takes to restore a piece of equipment to optimal working conditions following a breakdown.
- MTBF is the Mean Time Between Failures. This element predicts the time between one failure and the next failure. This can help predict the life span of the piece of equipment.
- Availability, which is expressed as a percentage, is a measure of how well a piece of equipment can operate as desired under reasonable conditions. It can be improved by undertaking appropriate maintenance activities. It is calculated by dividing MTBF by MTBF + MTTR.
- Wrench time is the time that maintenance personnel actually have a tool in hand and are working on the actual repair or maintenance activity. It does not include time waiting for parts, in meetings, or traveling to and from job sites. No one can have 100% wrench time. The lower the number, the less efficient the operation is, the higher the

number the better, more efficient it will be. Actions such as ensuring parts are available or can be ordered quickly, training in best practices, and having personal protective equipment readily available will help improve wrench time.

- Ratio of proactive work to reactive work. This is a measure of the efficiency of the organization.

An organization that has a robust work order system can analyze data from the system to assess each of these metrics and observe the trends over time. A dashboard could also be created to display these metrics. The system managers can make adjustments as necessary to improve the maintenance practices.

7.0 REPAIR, REHABILITATION, AND REPLACEMENT INTERVENTIONS

When an asset reaches a failed state, there are three basic options. It can be repaired (which is essentially corrective maintenance), be rehabilitated, or replaced. There are many factors that go into which decision is made including: whether there are repair and rehabilitation options for that asset, the cost of each option, how much the options extend the asset's life, whether the repair or rehabilitation options meet levels of service requirements, the underlying structural integrity of the original asset, the time it takes for each option, whether the utility has the capability of doing the work themselves, and the criticality of the asset. It is also possible that options may be mixed if the intent is to replace the asset, but the time it takes to do that is too long. In these cases, the asset may be repaired (if such an option is possible) to keep the asset in service while a rehabilitation or replacement is planned and executed. It is unlikely that an asset would undergo rehabilitation before replacement. Generally, rehabilitation is done instead of replacement.

Sometimes, although a repair can be made, it is not the best option. Consider the example in Figure 5.14. The picture on the top left is one small street in a neighborhood. All the other pictures are of pavement cuts on this small road over the course of the last couple of years, including one that shows a recent cut along with four other pavement cuts in the exact same area. In this case, the service lines are leaking because they are made of a problematic pipe type that fails frequently at the joints. While the pipes have been repaired numerous times, the effectiveness is questionable. The better approach would most likely be to replace the service lines rather than continuing the repairs. The system could remove and replace each service

FIGURE 5.14 Photos of Pipe Repairs

line when it leaks, rather than making a repair, reducing the number of future repairs over time.

Whenever an asset fails, especially high dollar and highly critical ones, it is important to consider a full range of potential response options. Some possible options are listed below:

- Implement Temporary or Permanent Bypass Solution: If the asset can be bypassed permanently without significant impact, this may be a reasonable option.
- A Non-Asset Solution: When the problem can be solved without any asset at all, it is called a non-asset solution. One example is implementing an aggressive water conservation program to reduce usage and prevent the need for a failed water source to be brought back online.
- Reconfigure and Implement New Process Solution: If there is a different way to provide the necessary function of the asset, reconfiguring and implementing a new process may be a good option. For example, if one type of treatment has been used to remove a contaminant but there are better options that have been developed since that time, it might be best to implement the new solution rather than repairing the old one.
- Long-Term Asset Repair: A repair option that can allow the asset to continue to operate for a long period of time without needing additional, frequent interventions.
- Rehabilitate Asset: Use a technique to bring the asset back to almost like-new status and greatly lengthen the useful life of the asset.

- Replace With Same or Similar Asset: Install the same asset or one very similar to it.
- Redesign (Replace With Another Asset): It can be worthwhile to replace with a different asset if there is a better option that will increase efficiency or improve customer levels of service.

Which option is chosen depends on several factors, such as whether a bypass option is available or possible, and how it would affect operations and whether it can be short or long-term in nature. If a non-asset solution is possible or the system can work without the asset using another process, the system can be reconfigured and a new process can be implemented. With a long-term repair, it is necessary to determine if a repair is possible and if it will provide a reasonable amount of additional asset life, the cost compared to rehabilitation and replacement, criticality of the asset, total down time, and whether the repair can be done in house or with a contractor. In rehabilitation, one of the biggest issues is whether there is a rehabilitation option. Other considerations are the criticality of the asset, cost, remaining life after rehab, and total down time. In replacement, there are two options—replace with the same or similar asset and redesign and replace with a different asset. Which one is selected depends on cost, criticality, total down time, reduced maintenance requirements, reduced repair potential, redundancy, ability to fix other issues, and energy and water efficiency improvements.

Decisions also relate to criticality of assets. What options are chosen can relate to where the asset is in terms of criticality. Those assets that are very high risk are likely to be rehabilitated or replaced rather than repaired. They will probably only be repaired if the underlying integrity of the asset can be maintained and if the repair will provide an extended life span. They may also be repaired if there is an interest in temporarily fixing the asset while waiting for a permanent replacement or rehabilitation. The intent with these assets is to replace or rehabilitate early to prevent the actual failure from occurring.

Figure 5.15 demonstrates how criticality relates to various options for failed assets. For assets in the upper left box, the assets should be monitored to determine if the condition is deteriorating and the probability of failure is increasing. When the probability of failure (POF) is high enough, the asset should be replaced, before a failure. The asset can be repaired when the asset life can be maintained. For assets in the lower right box, the assets will most likely be run to failure and repaired after the failure. These are called managed failures. These assets can be repaired until replacement or rehabilitation is required. For assets in the lower left, there should be little need for replacement. These assets can be repaired when necessary.

Consequence of Failure →

Monitor and then when higher POF, replace or rehabilitate Repair when asset life can be maintained	Replace or Rehabilitate Only repair if underlying integrity of the asset can be maintained or if it is necessary to keep the asset operating until replacement
Managed Failures Repair is a good option	Managed Failures Repair will be a good option until replacement is absolutely necessary Rehabilitation can be good option

Probability of Failure →

FIGURE 5.15 Repair/Replace/Rehabilitate and Criticality

A managed failure is one in which the asset is intentionally allowed to fail. This approach is only used on low-consequence assets because the asset will fail so the consequence will happen. If the consequence is not high, it is okay if it fails. With managed failures, it is necessary to have procedures in place for how the asset will be repaired after failure. Spare parts should be available as well as equipment, necessary supplies, and the staff should be well-trained in repairs. If an outside contractor is going to be used to make the repairs, the contractor must be available on call.

As an example, a large system thought it needed to replace all of the pipe in one section of their distribution system. All pipe was 75 years old, 6-inch cast iron pipe. Available information included a main line break database that had some useful information that almost no one knew about, as well as some old paper as-built maps, which created a picture of the system. Based on an analysis, only two pipes needed replacement, a few needed to be watched to see if they began to deteriorate faster while most of the rest were fine with either no breaks or 1 break in the past 15 years. This analysis prevented the system from spending unnecessarily on pipe replacement. Since none of the pipe was high criticality, it was possible to run most of the assets to failure.

8.0 CAPITAL PROJECTS

When it is time to replace assets, a capital project may be necessary. This type of project is one in which outside funding may be required to obtain the sufficient funding to complete the project. This outside funding may come from state or federal funding sources or other mechanisms, such as bonds or a commercial bank. The most important part of the process is making sure that the project is based on a data-informed process.

A good starting point for all capital projects is to ask the following questions:

1. What is the need for the project? What is the business purpose?
2. What is the benefit of doing the project?
3. How will customer levels of service be affected in the short and long term? Will they improve, stay the same, decrease initially and then level off or improve?
4. What is the risk if the project does not go forward? What negative effects might occur? When will they happen? To whom would they happen?
5. Does the project address any and all regulatory issues you are currently facing or expect to face in the future?
6. Does the project involve assets that are included in the priority ranking from the AMP and are they high-priority assets? Was this ranking based on data from the operations and maintenance and related work order data of the facility?

If these questions cannot be answered adequately, additional data should be gathered before moving on. Once the questions can be answered, the answers should be written down so that they can be expressed clearly to elected leaders, the general public, and potential funding sources. Ultimately, the customers will have to support the project no matter who the funder is, so it is important that the reason for the project, and particularly how customer service (as defined by the level of service goals) will benefit, can be clearly communicated. (It is important to remember that even in the case of grant funding, the customers are paying something as their tax dollars are going to support the governmental entity who is supplying the funding.)

A well-thought-out project is one that has a clear need and business purpose. There are distinct, expressed benefits of doing the project, levels of service benefits in some way (even if it is only to keep levels of service from declining if the project is not completed), there are adverse risks to the utility or customers if the project is not completed, and all regulatory issues, now and into the future are addressed. Furthermore, these answers are not just speculative; they are based on data from the facility operation and maintenance, hopefully through a robust asset management program that includes a work order system.

In the early stages of the project as these questions are being considered, there are three possible options:

Option 1: Project has a clear need and benefit, is consistent with strategic goals, addresses/improves service-level goals, addresses regulatory requirements

Option 2: Project need is not clearly identified, benefits uncertain, not consistent with strategic goals, no identification of impacts to levels of service goals or does not fully address regulatory requirements

Option 3: Project not well considered and no data backs up the need for the project. Strategic goals not addressed, customer service not improved (as expressed by level of service goals), regulatory issues not addressed

Options 2 and 3 will be discussed first, followed by Option 1. In Option 2, some elements of a good project are in place but others are missing. In this case, it is necessary to reconsider the project before proceeding. It may require rescoping the project to eliminate elements or add others. It might also be necessary to reconsider levels of service and how the potential project would improve service. Once the project can adequately address all the questions, it can proceed as Option 1. If it cannot address all the issues adequately, it should proceed to Option 3 and not go forward.

In the case of Option 3, the project elements are not well supported, and the project should not move forward. The biggest difference between Options 2 and 3 is that in Option 2, the project should be reconsidered to see if it should go forward and proceed to Option 1 but in Option 3, it is clear that the project should not proceed.

An Option 3 Case Study: A small utility in the southwestern U.S. had applied for funding for a project to replace all of the utility's distribution piping at a cost of $5 million. The project was not backed up by data; there was a general "the pipe is breaking a lot and costing too much money." This particular utility was ready to move forward with the project when it learned about asset management and began considering the project with an asset management mindset. The review of any available data to determine whether this was an Option 1, 2, or 3 project became the first priority. It was determined that the best available data set to calculate the number of pipe breaks as well as the cost to repair each break was the utility's board meeting minutes. Each month, the contract operator would attend the board meetings and state where breaks occurred and the cost to fix each break. In general, the break repair costs were a set rate. Examining the monthly meeting minutes revealed that the overall break rate was low and it was not increasing over time. Essentially, the data did not support the project going forward and it would not have improved levels of service in any meaningful way while costing customers a considerable amount (having to pay debt service each month on a $5 million loan). It turns out that there were some actions that could be taken to improve levels of service for the customers but it did not require a capital project. In the end, this project was not completed as initially scoped because it was not possible to answer the questions in a way that backed up the need for the project. Instead, a much smaller, less expensive project was completed that included adding some loop lines, some isolation valves, and a few other necessary changes. The project cost $50,000, saving $4.95 million, and improving level of service by improving water quality at the dead ends and reducing the number of customers that had to be shut off to make pipe repairs on the main transmission main.

How would the previous example answer the questions?

1. What is the need for the project? What is the business purpose? The professed need was to reduce breaks, but the data showed that there was no breakage problem. Therefore, there is no real need for the project as proposed.
2. What is the benefit of doing the project? The purported benefit was to have fewer breaks to repair, but this benefit was very low and not worth the cost.
3. How will level of service goals be affected in the short and long term? No impact. There were customer service–related issues (such as poor water quality) that were not going to be addressed with the project because long dead ends would still exist. During construction the customers would incur a negative effect without receiving a positive effect after construction.
4. What is the risk if the project does not go forward? What negative effect might occur? When will they happen? To whom would they happen? Little to no risk. There was one area on the pipe that required a more significant repair. In that one area, there was the potential for some continued repair work if nothing was done. This negative effect could be mitigated in a much simpler, less expensive way than doing the whole project.
5. Does the project address any and all regulatory issues you are currently facing or expect to face in the future? There were no regulatory issues to address then or in the near future. However, there were some water quality issues related to long water age that the original project was not going to address. Those issues were initially being addressed by frequent line flushing at the dead ends. The smaller project that was completed addressed these issues and eliminated the need for frequent flushing.

The most important aspect to the example is understanding that some projects simply should not be done. There is no reason to spend the money in the manner suggested. If the project can be re-scoped into something reasonable, it can proceed in that manner.

Thus far, Options 2 and 3 have been discussed. With projects in Option 1, if the project answers the questions adequately indicating that further consideration should go forward, the next step is to consider the alternatives regarding how the problem can be resolved. It is also necessary to consider a non-asset solution if there is the opportunity for such an option. An example of a non-asset solution is deciding to give employees credit cards to use at a commercial gas station rather than repairing or replacing an existing municipal run gas station.

The alternatives analysis was discussed above in section 3.0, Blending Design and Operation, along with how important it is to bring utility personnel into the alternatives process. Once the preferred alternative is selected, it is necessary to determine the estimated project costs, both the initial construction costs and the overall life cycle costs for the entire life of the project. Given that the project is being completed to provide some benefit to the community, it is necessary to determine if the benefit of the project outweighs the cost. If the benefits are greater than the cost, it should proceed. If the benefits are much less than the cost, it is necessary to take another look at the project itself and the selected alternatives to determine if there is a less expensive way to achieve the benefits. If the costs are close to the benefits, then further analysis may be warranted to determine if this is the best course of action. If the project is to move forward, the next step is to secure the funding. Funding for capital projects will be discussed in the next chapter, Long-Term Funding.

Ultimately, the goal is to determine if you are doing the right project at the right time for the right amount of money. Because the project will need to be approved by the elected leaders and be acceptable to the general public, it is necessary to ensure that the project meets these criteria. In some cases, even if the benefits outweigh the costs, it may not be possible to do the whole project at once. In these cases, the project may need to be phased to allow portions of the project to be done before others. If phasing is undertaken, it is vital to have the right portions of the project in the first phase. Highly critical assets need to be included in the first phase to prevent a catastrophic failure from happening during or soon after a new construction project. If this type of event did occur, the customers may lose confidence in the utility and the utility may not have sufficient funds to complete the repair or replacement of the failed asset.

Case Study: Small utility in the southwestern U.S.: The utility was looking to install new water meters using radio read technology so that operators would not be required to read the meters manually. However, when an asset management improvement initiative was advanced and the criticality portion of the process was done, it was clear that there were assets that were much higher priority than the meters. In fact, a pump was near failure and it was the only pump supplying water to a big section of the community. If that pump failed, a portion of the community would be out of water for a considerable period of time. The water system personnel were initially considering only their inconvenience and not the true criticality of assets. Once the criticality analysis was done, the project included the pump and a few other assets and the meters were put off. Eventually, the utility was able to get funding to do the meters a few years later.

9.0 REFERENCES

The Deming Institute. (2025). *Plan-do-study-act cycle.* https://deming.org/explore/pdsa/

Gibson Jr., G. E., & Hamilton, M. R. (1994, November). *Analysis of pre-project planning effort and success variables for capital facility projects* (Source Document 105). Construction Industry Institute, The University of Texas at Austin.

Ledet, W. P., Ledet, W. J., & Abshire, S. M. (2012). *Don't just fix it, improve it!: A journey to the precision domain.* ReliabilityWeb.com.

Chapter 6

Long-Term Funding

In the long-term funding portion of asset management, it is necessary to consider the amount of money the utility has to cover all of its operating expenses (op ex), capital expenses (cap ex), and debt service, whether this amount of funding is sustainable, and whether it is sufficient to cover the required expenses. This component also considers the potential sources of money that are available for addressing operating and capital expenses. Operating expenses are generally paid for out of revenue from user charges of some type or form while capital expenses can come from user fees or outside sources—state, federal, bonds, or private funding. When outside sources are used, they often need to be paid back over time, unless they were grant funds. In these cases, the user charges need to include sufficient revenue to repay the loan funds or bonds.

Long-term funding recognizes that money is necessary at every stage of the asset's life to reduce failures, maintain or improve levels of service, extend asset life, and ensure long-term sustainability. It is important to understand the need for funding as well as secure the necessary revenue to have a successful utility.

1.0 FUNDING AND TIES TO OTHER PARTS OF ASSET MANAGEMENT

No utility ever has *all* the money it needs to do everything it wants—to meet all of the customers' expectations, to do all routine and preventative maintenance tasks that are required, to gather all the necessary data, to perform all the predictive maintenance and condition assessment that could be done, to replace all assets before failure occurs, and to meet every regulatory requirement. Utilities must constantly make choices regarding where to spend their money. If they spend it on A they cannot spend it on B and C. Or maybe they choose to do some of A, B, and C, but not all aspects of any of them. There is always a gap between the money we need to do all the things we want and the money we actually have.

The first four components of asset management are intended to make the utility more efficient and effective and to reduce overall life cycle costs. These parts help reduce the cost of doing everything the utility wants to do by focusing on the highest needs, eliminating unnecessary tasks, optimizing operations and maintenance, and focusing on level of service goals. This last component of asset management is focused on better understanding the need for funding and attempts to increase the amount of money available to perform necessary tasks. The work to improve efficiency and increase funding reduces the gap between the money we want and the money we have.

The more clearly utility staff can identify the amount of money needed, what it will be used for, benefits to be derived, and the consequences of inadequate funding, the more likely the governing body is to understand the need for increased funding and to support it. If customers can be made aware of the purpose of rate increases and understand that the increased rates result in increased or sustained service, they may be willing to support them. While they may not want to pay higher rates, if they understand and accept the reasons, they may not speak out against them.

Practicing asset management does not guarantee that elected leaders will make the right decision regarding operational budgets, funding for replacing assets, or the need to reduce overall business risk. The leaders/decision-makers are ultimately responsible and make whatever decisions they deem best. However, asset management has the potential to increase the ability to make a better case, especially data-driven cases, for funding and to help decision-makers understand the benefits of the funding or consequences of not increasing revenue. Once the decision is made, the utility personnel must do the best they can within the given budget to ensure that negative consequences do not occur but should recognize that it may not be possible to avoid all negative consequences. If/when the negative consequences occur, the utility personnel should seek guidance from the decision-makers on how best to deal with and communicate about the events.

The job of staff (management and operations) is to clearly communicate all information and allow decision-makers to make the most informed decision possible. If information is withheld, the decisions may not be in the best interest of the customers. As an example, if the staff believes that three pump station upgrades are required to keep the wastewater collection system operating and handle the current and future flows, they must tell the decision-makers. They cannot assume that since there are currently insufficient revenues available, they will only request funding for one pump station instead of three. The decision of whether there are sufficient dollars must be made by the decision-makers. If they are only told about one pump station, they cannot make the decision to raise rates to cover the costs of all three, if that is what seems most desirable for the system. If they choose to fund one or two, the utility personnel should use asset management thinking to determine the highest risk pump station(s) and replace that one(s) first. Then it should do its best with maintenance and repairs to keep the remaining station(s) operational until funding can be secured.

2.0 FUNDING THE FULL COSTS OF OPERATION

When running a utility, it is necessary to fund the full costs of operation to be sustainable over the long term and to continuously meet customer expectations. There is a bit of debate over exactly what this terminology means in practice, however. It is clear that user fees should cover operations, including salaries, treatment chemical costs, electricity and other utilities, land and facility costs, and insurances. Similarly, the costs related to routine and preventative maintenance including supplies and equipment as well as costs for any outside contractor must be covered.

The costs of corrective maintenance, such as repair costs, including the spare parts needed to make the repairs, are also part of full-cost pricing. Additionally, spare parts that should be purchased for future repair/replacement, especially those related to critical assets or that have long lead times, should be included.

How much predictive maintenance to fund is a bit trickier question. A fully funded operation may include a lot or a little predictive maintenance. This type of maintenance can be cost-effective (meaning the benefits outweigh the expenses) for highly critical and some moderately critical assets and should be included at that level. Not every asset that can potentially have predictive maintenance should receive it because for low criticality assets in which the costs of the predictive technology are high, the costs may outweigh any benefits.

Some amount of asset replacement cost should be included within full-cost pricing, especially for short-lived assets, lower-cost assets, frequently replaced assets, or those that need to be replaced more quickly than can be accomplished with outside funding sources. However, it is difficult to include the full cost of replacing an entire water or wastewater system because that can amount to hundreds of millions of dollars and saving that much money in an account (such as a reserve account) can be problematic. But it is still worthwhile to determine an amount of money that should be set aside (where allowed by local and/or state regulations/rules/laws) to enable replacement of assets.

Another way to think about funding replacements is to fund depreciation. Assets depreciate over time, which means they lose value. If the lost value is set aside in an account, such as a reserve account, each year, by the time the asset is fully depreciated there should be sufficient funding to replace it. This methodology will result in collecting significant additional revenue and will result in a large sum of money in an account.

One concern with a large reserve account to fully or mostly pay for a new system is that the account may be accessed for other purposes rather than the new assets for the water facility, to the extent that this practice is allowed by state or local rules and regulations. While this may seem like a completely unreasonable action, the other side is that the water facilities are part of a community but not the entire community. In many communities where water and sewer are a component of municipal services, water needs could be competing with many other needs—schools, roads, libraries, pedestrian pathways, recreation facilities, emergency services (police, fire), health care—and those needs may at times greatly outweigh the needs of water. If these needs become urgent, a community's leaders may find ways to take the water money and use it for other purposes. (This situation occurred in some jurisdictions after the great recession of 2008.)

While water funds may be accessed for other municipal purposes, sometimes the reverse is true. The water utility charges rates that are too low to adequately fund the full operations, so transfers are made from the municipal general fund to the water fund to cover necessary expenditures. This situation should be considered a short-term solution and action should be taken to raise revenues such that a general fund transfer is not required. Furthermore, general fund transfers may not be equitable if all members of the municipality or governmental entity are not water customers. For example, while all members of the community pay property or sales taxes or other revenues that make up the general fund, some may be on private wells or septic systems. The intent should be to run the drinking water and wastewater systems as enterprise funds in which the revenues generated are sufficient to cover all costs, plus reserves, and funds are neither taken from nor input into the enterprise fund.

Another issue with fully funding future replacement is the idea of equity. Should today's residents be paying for long-lived assets that will benefit future generations? For example, if sewer and water pipe last 75 years but the residents pay for it over 40 years, have they been inequitably asked to pay for future users? This issue needs to be considered when deciding how much replacement to fund out of rate revenue.

If the utility chooses to fund new assets with outside funding, which may or may not meet the definition of fully funding the utility, the debt service must be covered as part of the full costs. If the utility receives grant funding, it does not have to pay the funds back but does receive the full benefit of having the assets. In these cases, is the utility fully funding its operations?

Finally, the utility can choose to fund reserve accounts each year. These reserves are discussed in the section on budgets. The reserve accounts provide additional funds for the utility if there are budget shortfalls, to ensure payment of debt, to cover emergencies, and the like.

Each utility must decide, beyond the basics, what it means to fully fund its operations. Once that is determined, the revenue must be set high enough to cover those expenses.

3.0 USER RATES AND FEES

The vast majority of the utility's revenue to fund its operations, maintenance, and some repairs, rehabilitation, and replacement is likely to come in the form of user rates and fees. In the case of drinking water, there is often a base bill (a minimum charge just to have service) as well as a volumetric charge (a fee based on the quantity of water used). In wastewater, there may be a flat fee or a base bill and volumetric charge where the volumetric charge is based on drinking water usage. In arid climates, usage varies greatly from winter to summer and the extra summer usage is associated with outside watering that does not end up in the sewer system. In these cases, the sewer charge may be based on winter average usage rather than monthly usage throughout the year. In other cases, wastewater may charge for a portion of the water usage all year (say 80% of water usage) to account for the portion of drinking water that does not return to the sewer (usually outside water uses). In some cases, wastewater is charged as just a flat fee due to the complexities of gathering the water usage data and adjusting it for wastewater flows or in cases where a large portion of customers are on private wells. For stormwater, there may be no fee at all, a flat fee, or a fee based on impervious surface area or other means of assessing stormwater flows to customers.

There is no single right way to set rates. Rate design is more of an art than a science. There are many different ways to achieve the same total desired revenue. It depends on what the goals of the utility are and what it is trying to achieve. Goals that are often associated with rate setting include: full cost recovery/revenue stability, encouraging conservation, fostering business-friendly practices, and maintaining affordability. It is not possible to maximize the rates to achieve all of these goals as some of them will be mutually exclusive with others. As an example, a utility could choose to charge a flat fee. In this discussion, we will take it as a given that this fee is high enough to generate the needed revenue for the utility. This type of rate is very good for revenue stability. The utility would know exactly how much revenue will be generated on a monthly basis because revenue is completely independent of usage. This type of rate structure would be easy to budget for. However, this rate structure is not conducive to water conservation. The users will pay the same amount whether they use a lot of water or a little, so why conserve? Furthermore, the users will have no record on their monthly bill that indicates how much water they have used and whether that has increased or decreased from the month before. Without this information, it is extremely difficult to moderate household usage. Most individuals are not capable of estimating how much water they use on their own without a meter. It is hard to estimate how much water is used to water a garden or the lawn or children playing outside with a hose. A flat fee for all is also not equitable. Every user is paying the same even though some are using significantly more water than others. Therefore, while this rate is good for revenue stability, it is not well-suited to encouraging conservation or addressing affordability. It may or may not address business-friendly practices.

Rates that include a variable charge are better for conservation and can include a large price signal to discourage high water usage. They can also help the utility manage its water losses. Without customer usage information from accurate meter reading, it is not possible to do any more than guess at overall system non-revenue water (the amount of water the utility produces but receives no revenue for). Non-revenue water includes real water losses as well as authorized unbilled consumption and apparent or paper losses (inaccurate meter readings, water theft, data handling errors). Rate structures that include variable charges may also be valuable for low-income customers, depending on how they are structured.

Business-friendly practices typically include charging less for more water used. This type of rate is called a declining block rate and while it may be positive for attracting businesses that use large amounts of water, it is probably not favorable to affordability (lower water-using customers will be paying for the additional cost of the higher water-using customers) and is not a water conservation strategy. When a user sees a reduction in their per gallon charge while using more water rather than an increase, there is less incentive to reduce water usage. Another business-friendly practice is to charge less for a business than for residential customers or to tie rate increases for businesses to rate increases for residential customers. Similarly, either of these practices is likely to shift the burden of paying for the utility from businesses to residential customers. If this type of rate structure is practiced, it should be transparent to the community so that they understand the choice that the utility is making to try to attract business to the community.

A medium-sized utility in the southwest decided that it wanted to incentivize business in its community in part through its water rates. The utility decided that the businesses would not have a higher rate than residential customers and that their rates could not be raised at a higher percentage than residential customers. Several years later, the utility determined that the businesses were putting a much higher strain on the water and wastewater facilities and that the businesses were not employing very many people from that town; most employees were from neighboring towns. Unfortunately, they were unable to change the rate structure for a long period of time and residents had to subsidize the businesses' water usage.

Because rates can be structured in so many different ways it is necessary for the utility to choose what its goals are and design the rates accordingly. Rates should be reviewed annually to make sure the financial needs of the utility are being met, and they should be raised periodically, preferably annually. Sometimes rates are set to incrementally increase based on standard factors, such as the consumer price index (CPI). While an annual increase is applauded, it needs to be understood that the CPI is not related to water production, wastewater treatment, or stormwater handling. It is unlikely that indexing on the CPI will accurately reflect the true increases in costs for the utility. A better approach is to determine the increase in costs for the utility and increase rates accordingly. However, because rates have a political element in most cases (elected leaders need to make the decision about rate increases), it is easier to index on a factor than to raise rates using an annual vote of the elected leaders. Alternatively, some utilities choose a standard percentage increase each year. This may be better than the CPI or other index but shares some of the same issues in that it is not necessarily going to match the actual increase in costs.

A further discussion of rate setting is considered beyond the scope of this book. There are many good resources available to learn more about rate setting and how it can best meet the needs of the utility and there are many consultants who can help a utility think through these issues. The main takeaway for the relationship between rates and asset management is that rates will likely be the primary method for funding the utility and if the rates are inadequate, management of the assets will suffer.

4.0 MAKING THE CASE FOR INCREASED RATES TO CUSTOMERS

Making the case for increased rates involves considering the best way to reach customers in the specific community and helping them understand the purpose of and need for additional revenues. This is a chance to be creative in terms of how to reach customers and to consider using multiple approaches. The more customers understand the services they are receiving from the water facilities and how the additional revenue will ensure continued service or increase service, the more likely they are to agree to rate increases or other changes in fees. Several potential options for customer outreach are presented here, but these are by no means the only ones.

Customer Opinion Surveys: One option is to conduct a customer opinion survey to find out what customers are thinking and what level or service they would like. This information can be used to help determine needed outreach and information sharing and how best to reach customers. The customer opinion surveys should contact random customers and use both landline and cell phones. This approach is a bit more complicated if the utility does not collect customer phone numbers as part of its customer information. If that is the case, the utility may have to conduct a mail-based survey instead.

Customer Conversation Meetings: There can be customer conversation meetings in which participants are recruited through websites, emails, social media or any other methods that work for that community. The meetings should be interactive and engaging and governing body members should be encouraged to attend so they can hear directly from customers and interact with them. One utility that held customer conversation meetings was able to determine that the customer priorities for the utility were significantly different from those of the employees and leadership of the utility. That was important information for the system to find out. Afterwards they were able to adopt practices that would help better explain the utility's priorities and why these were the priorities. They were also able to adjust some of their activities to include customer interests.

Traditional Media: Media should also be used to the system's advantage. Utilities can work with the local media to get their message out. This can include print, TV, and radio media sources. With proper information, it is possible that the media may provide supportive editorials for the rate increases.

Additional Methods: Some other options can include bill inserts, billboards, large signs, public service announcements, and tours. Tours can be a very effective way for customers to have tangible information regarding the complexity of drinking water and wastewater treatment and interact with water utility staff. Sometimes the tours are offered for the general public and other times for school-age children. Billboards can be used in locations that allow billboards (not all states or jurisdictions permit billboards) to present information important to convey, such as water conservation, preventing fats, oils, and grease from entering the treatment plant, or construction projects. Large signs can be posted near construction projects to convey information about the nature of the project, who is paying for it, and the time frame.

Websites and Social Media: Information can also be provided on websites or through social media sites. Given the explosion of information on the internet, all utilities should consider having a website, either stand-alone or as part of the larger community's website. Customers who are looking for information about the utility will search the internet as a first step. The more information available, the more likely customers will find what they need from the utility's source rather than somewhere else. At a minimum, the utility should share basic information, rate information, public meeting notices, and contact information. The website is also the perfect place to share asset management related information, including the mission statement, external level of service goals as well as progress towards meeting the goals, and the asset management plan.

Social media is also an extremely popular way of sharing information and there are many options available. The social media platforms used by the utility to share information with customers should be ones that the utility feels comfortable with and that have the best chance of reaching customers. While social media can be an opportunity to get positive messages out, some customers may post negative stories, including stories that may be completely unfounded. The utility should be ready for this type of event if they decide to use social media as a main method of communicating with customers. Utilities may also have to moderate their social media sites or websites to ensure that no inappropriate content is posted.

Funds for Specific Purposes/Projects: Another potential means of getting support for rate increases is to create a special fund to enable the utility to address a very specific need. This might be a specific construction project, one type of asset, source water protection, or removing a secondary contaminant. In one example, a facility in the southern United States set up a fund for source water protection. They determined that protecting the lake that was a main source of water supply would have significant cost benefits over time. The lake was facing considerable development pressures and the utility needed to purchase land around the lake or at least conservation easements. By thoroughly describing the purpose of the special fund, the utility was able to add a specific fee on customer bills that could be used for source water protection, including land purchases. There was no outcry in the community, and the project was very successful.

Another utility in the south created a similar fund to support the replacement and rehabilitation of a specific type of asset that was causing considerable problems. They were able to gain enough money to make all the fixes for that type of asset. The fund was then put towards a different set of assets. It may be easier to get support for rate increases when customers are well-informed regarding the need for and use of funds. In these cases, it is also possible to clearly tie the money to an outcome. Customers can see the land purchases for example or observe the construction taking place. They may experience fewer outages of water or have higher aesthetic quality of water entering their home.

5.0 OTHER FORMS OF REVENUE

While rates and user fees are the most common method of obtaining revenue, there are other fees and forms of revenue. Some utilities will use taxes, such as sales tax, to provide revenue to the water system. There can also be impact fees (which can be called by other names) that require a developer to pay a utility a fee that covers the increased costs associated with providing service to the development. There can be penalties associated with late payments or for shutting off and turning on water for lack of payment. It is important that these penalties cover the full cost (and hopefully a bit extra) for the utility to perform the service. In many cases, especially in the case of very small systems, the payments are too low. Some utilities charge a fee that is set based on what the management thinks can be sustained rather than by what it costs to shut the customer off. As an example, one utility charged $15 per shut off when the true cost of performing the service was closer to $50. One consideration that is very important is that shut offs must be equitable and fair. It is not acceptable for utilities to ignore higher income customers who fail to pay and shut off lower-income customers much more rapidly. Everyone must be treated fairly.

Revenues can also come in other forms, such as payments for cell phone companies to rent water tower space. These revenues may have fewer regulatory restrictions than general rate revenue so they may provide more flexibility in how they can be used.

When setting rates, a utility can also consider customer assistance programs to help low-income or low-resourced customers pay their bills. In some cases, these programs will be built from internal funds from the utility or community (according to state law), while in other cases funds can come from outside the utility service area. These funds may be philanthropic dollars or provided by a state or federal program. While this type of program does not bring in additional funding to the utility, it changes the source of that funding. If a utility is interested in setting up a customer assistance program there are many sources of information to investigate. For further information about water affordability, two references are: U.S. Environmental Protection Agency's (EPA's) 2024 report titled, "Water Affordability Needs Assessment: Report to Congress" (EPA, 2024) and the Environmental Financial Advisory Board's (EFAB's) 2025 report titled, "Advancing Water Affordability Nationwide: A Framework for Action" (EFAB, 2025).

6.0 BUDGETS

Similarly to rate setting, a thorough discussion of budgeting is also outside the scope of this book. As with rates, there are many good sources of information on budgeting that can be consulted to determine how to create and use a budget. Within asset management, there are a few areas of budgeting that are worth considering so those aspects will be presented in this section.

A budget is likely to be required for many entities, especially municipal utilities, but whether it is required by law or not, a budget should be developed for *every* utility—drinking water, wastewater, and stormwater. The budget indicates the expected level of expenditure in every facet of the utility as well as the expected revenues and is an opportunity to appropriately plan for the highest priority needs. The budget should be approved and formally adopted by the governing body to ensure that everyone sees it as an effective means of properly managing the utility and its financing.

One aspect of budgeting that ties very much to asset management thinking is that it is much better to set a budget based on what is *actually needed* rather than based on what you think you can get. Utilities may be artificially hampering their ability to get the proper funding because they do not want to share the hard truth with upper management or elected leaders. It is necessary to determine what amount of money is needed for assets to be managed sustainably and effectively while meeting customer expectations. A utility costs money to run and there is no escaping this fact. The truth may be hard to convey but it is a necessary activity.

Developing the budget should involve all aspects of the organization including operations and maintenance as their needs will be crucial to successful asset management and efficient and effective operations. The budget can be based on the prior year's expenditures with adjustments for any expected changes (either increases or decreases) or the budget can be zero-based, meaning that each year the budget is built from scratch with every line item starting at zero and building up to the appropriate amount. Starting from the previous year's budget is much easier, but runs the risk of encouraging individuals to spend all of their budget—whether they need it or not—out of fear that the budget will be cut the next year if they do not. It might also make it harder to determine potential changes from year to year in the budget categories. A zero-based budget provides an opportunity to create the budget based on more exact needs, but it is time consuming to prepare. In either case, information that provides a thorough understanding of the prior year's expenditures is a necessary element of the process. With a robust asset management or computerized maintenance management system (CMMS) software program, this data should be easy to acquire.

In addition to the upcoming year's expected expenditures, it is necessary to forecast not only following year but also several years (5 years total is a good time frame to consider) of anticipated expenses. Here again, the asset management system can be helpful. Based on the condition of assets, the probability of failure, and past information on asset downtime, predictions can be made regarding asset repair and replacement. If there are small-dollar assets that are expected to fail over the next year (i.e., asset replacements that can be built into the budget rather than needing outside funding), the replacement costs can be added to the budget. If spare parts are needed to build resilience or to replace parts used over the previous year, the asset management (AM) software may also be able to provide this information.

The expense budget must be compared to the expected revenues for the upcoming year. If the necessary expenditures will outpace expected revenues, either the budget must be reduced or the revenues must be increased. If the budget is reduced, the consequences of this decision must be understood. Will there be serious reductions in service levels? Will some assets with high consequences be expected to fail? Will some asset replacement have to be put off for a while? Will a major capital project have to be phased or put off? If the consequences are too great, a revenue increase may be the only answer.

Once the budget is developed and adopted for the next year, it is necessary to have a process to compare the budgeted amount with actual expenditures on a monthly basis. It is especially important from an asset management perspective to compare the actual and budgeted amounts for operation and maintenance line items. If the actuals are above or below expected levels, further investigation is warranted to determine if the reasons are justified. For example, if the actuals are much higher, are there conditions that will reduce expenditures later in the year? If not, are there budget categories that can be moved around to make up for the shortfall? If the actuals are below expected levels, have activities been put off unnecessarily or to the detriment of the assets? Should the level of expenses be increased to ensure the assets are being well-managed? The actuals and budgeted amounts should be shared with the governing body monthly or quarterly (depending on how often the governing body meets). If actuals continue to be below expected amounts and activities are not being put off, an adjustment to those budgeted line items may be warranted.

Another way to potentially fill in gaps between expenditures and budgeted amounts is through reserve accounts. Unless prohibited by state or local regulations, every utility should consider having reserve accounts that can cover particular types of expenditures, especially emergencies. Types of reserve accounts can include: Operating Cash Reserves, Emergency Reserves, Planned Repair/Replacement Reserves, Planned Capital Improvement Reserves, and Debt Service Reserves. Whether a utility has all of these or just some of these is a decision for the governing body and the financial personnel. However, some situations may require reserve accounts, such as a debt service reserve when a federal loan is acquired. There are suggested amounts that should be in reserve accounts and one version of those suggested amounts is shown on Table 6.1 (Wichita State University & Southwest Environmental Finance Center, 2015). Other recommendations can be found in various guidance documents. The amounts in the reserve accounts should either meet requirements of the utility or balance the need for dollars against the need to be cautious about the size of the reserves. It is important to remember that customers are making decisions about how to spend their own dollars between food, housing, education, healthcare, utilities, and other necessary expenses so it is important not to overcharge and build excessively large cash reserves.

7.0 CAPITAL IMPROVEMENT PLANNING

Utilities need to develop a capital improvement plan (CIP) to detail the assets that must be replaced. The components of a CIP include:

- A list of required capital projects, major equipment, and any major studies
- A prioritized ranking of projects
- A timetable for construction or completion of the projects
- Justifications for the projects
- A financing plan

TABLE 6.1 Suggested Reserve Account Levels

Reserve Account	Purpose	Reserve Goal	Reserve Minimum
Operating Cash Reserve	Provide a cushion for day-to-day expenses	3–6 months' worth of operating and preventative maintenance expenses	2 months' worth of operating and preventative maintenance expenses
Unscheduled/ Reactive Maintenance "Emergency" Reserves	Provide funds for unforeseen repairs or replacement	Average expenditure for the past 5 years' unscheduled/reactive repairs PLUS an additional 10% of that figure PLUS the current cost of the most expensive capital item that is not in the spare parts inventory	An amount equal to the most recent typical year's expenditure for unscheduled/reactive repairs PLUS, the greater of (1) an additional 5% of that figure OR (2) the current cost of the most expensive capital item that is not in the spare parts inventory
Planned Repair/ Replacement Reserve	Purchase, repair, or rehabilitate items with typically a 1–10 year life span	The amount will be guided by the future asset repair and replacement costs predicted as part of the asset management program or the repair and replacement schedule	Without an asset management or repair/ replacement schedule, reserve an amount equal to 5% of the total system replacement cost OR reserve an amount equal to 3 months of operating expenses
Planned Capital Improvements Reserve	Construct or upgrade facilities in response to growth or change, including new regulations. Minimize debt accumulation. Provide matching or leveraging funds to help get grants or loans. Pay for pre-construction costs such as planning, environmental assessments or preliminary engineering reports.	The amount will be guided by the future capital improvement costs that have been predicted as part of the asset management program or capital improvement plan	10%–30% of the future cost of the anticipated capital cost
Debt Service Reserve	Ensure funds are available to meet debt repayment terms	Minimum debt service reserve or debt service coverage ratio will likely be specified by a loan or bond agency	Level required by debt covenants or funding agency

Source: Wichita State University & Southwest Environmental Finance Center, 2015

The detail included within the CIP will depend on the time frame of the CIP and how soon the individual projects will be completed. The sooner the project is expected, the more detailed the CIP information. The further out the project will be completed, the less detailed the information.

There is a need for both short- and long-term CIPs. Short-term CIPs are more specific regarding costs, project prioritization, and project justification. These plans are usually 5 to 10 years in length. In the case of 10-year CIPs, the cost information will be more specific for the first 5 years of projects and less so for the last 5 years. In both cases, the most detailed information will be provided for the items in the first 1 to 3 years. Short-term CIPs can be used for budgeting and seeking capital funding while long-term CIPs should be considered planning-level documents.

The long-term CIP should be created to cover from the end of the short-term CIP for up to 50 to 100 years. This CIP is rough costs only, does not prioritize the projects, does not need to include project justifications or a Business Case Evaluation (discussed later in this chapter). This long-term CIP is meant for planning purposes. It should help show customers and elected leaders that there are needs for asset replacement essentially forever. The system assets need to be renewed constantly and there will always be some projects that occur after the initial 5- or 10-year period of the short-term CIP. The better this need for asset replacements over the long-term can be understood, the more likely the elected leaders will be able to establish an ongoing capital improvement program. The long-term plan will show higher and lower values from year to year that can be averaged across the entire time period to determine an average annual incremental investment. This can help smooth out the major investment spikes.

One reason to have robust asset management software, or at least a robust asset management program, is to obtain the information you need to create your short- and long-term CIPs. Information such as risk, remaining useful life, and replacement cost can come from the software, or at least a portion of this information can be obtained.

It is important to understand that risk feeds into CIP planning, both short- and long-term. The assets in the high-risk category (high probability of failure [POF] and high consequence of failure [COF]) feed into the short-term CIP. These assets will need to be replaced soon, likely within the next 1 to 10 years. These are assets that the system wants to ensure are replaced prior to failure.

The assets in the moderate risk category, those that are currently low in POF, but high in COF, are likely to deteriorate over time and end up in the higher risk category. The deterioration may occur somewhere in the 10 to 50 year time frame, depending on the overall expected life of the assets and the rapidity of deterioration. These assets will show up on the earlier to middle portion of the long-term CIP.

The assets in the moderate risk category, those that are currently moderate to high in POF, but low in COF, are likely to fail sooner than the other moderate category or low-risk category.

These assets are typically in the managed failure mode. These assets are likely to fail within a 1 to 20 year time frame, but may not be specifically included on the CIP as they will be run to failure. Rather, the budget includes a line item to make repairs or replacements to failed assets based on the number expected to fail each year without specifying exactly which asset it will be. Because these assets are low COF, it is not cost-effective to spend considerable dollars assessing each one in this category to determine which is likely to fail and replace before failure. It is more cost-effective to allow the assets to run to failure and replace when failure occurs. If more assets fail than anticipated in the budget, it may be possible to handle those additional assets by shifting funding around within the operating budget or to use emergency repair funds. In some cases, assets in this category may be initially repaired to buy time before replacing them. In these cases, the specific asset will show up on the CIP. The assets in this risk category are expected to make up the earlier to middle portion of the long-term CIP, with some on the short-term CIP.

The assets in the lower risk category (those that are low in POF and low in COF) are those that are likely to make up the very long-term CIP projects, generally in the 30- to 100-year time frame.

For all high-dollar projects on the short-term CIP, a Business Case Evaluation or BCE should be completed. What constitutes high dollar depends on the utility. For one utility, this may be $100,000, for another $1,000,000 and still another $50 million. It should be those projects for which the utility will be investing a lot of money and those that may cause rate increases for customers.

The business case evaluation (BCE) starts with defining the drivers for the project. Why is the project needed? What are the benefits of doing the project and what are the risks if it is not completed? The BCE should also include a problem statement. What problem is the utility attempting to solve that requires an intervention? If the problem cannot be stated clearly, such that anyone, including elected leaders who may not be well-versed in water issues or the customer base, can understand it, the statement should be reworded until it is clear to all. The business case is not intended to justify a particular solution, but rather to be a holistic examination of a business need and the alternatives that could be used to address that need, including any non-structural (non-asset) solutions that could resolve the issue without additional infrastructure.

A non-structural or non-asset solution is one in which the problem or issue can be addressed without the need for new or rehabilitated assets.

The non-asset solution might be an operational fix, changes in customer practices, changes to utility practices, or elimination of a problem asset. As an example, one large utility owned and operated a fuel facility for its fleet vehicles. The fueling station required significant and costly upgrades. The utility decided that instead of investing in that infrastructure, it would provide its employees with purchasing cards that could be used at gas stations located throughout the community. Another example is a small community that was not meeting compliance. It thought it needed a brand new treatment facility but it was determined that operational changes could address the compliance problem without requiring new infrastructure.

The BCE needs to include a discussion of alternatives. Any project has alternatives. Some will be better than others for various reasons, such as price, ease of construction, minimizing inconvenience to customers, or technology that best fits the utility, but all should be examined. Part of the BCE process is to formulate the set of alternatives that will be evaluated. If there are some that are completely impossible, they can be removed at this stage, but all other reasonable alternatives, including the no-action or status quo alternative should be left in. The no-action/status quo alternative is meant to provide a comparison for other alternatives. If they are not better than doing nothing, the project may not be needed and should probably not proceed. If there are any potential non-asset solutions these should also be examined.

The BCE should not be considered a justification for a particular chosen project, but rather, it needs to objectively state the problem that needs to be evaluated, examine the holistic set of potential alternatives that can address the problem, including any non-structural or non-asset solutions, and provide sufficient information to allow a well-informed decision. The BCE process can be used to identify projects that should not proceed, projects for which another alternative should be chosen, cases where a non-asset solution can be applied, and cases in which the overall scope of the project should be narrowed. The use of the BCE has the potential to save considerable dollars within the capital budget.

Whether a project undergoes a full BCE or not, it is important to justify projects at least in a basic way and to examine alternatives. One medium-sized utility would go to its board with CIP projects and state each time that there were no alternatives to what they wanted to do. The board believed them and would approve projects without any additional scrutiny. This utility was in bad financial shape due in large part to the inability to think more creatively. As one example, they were looking at adding a new wellfield and stated that the only option was drilling new wells at one specific location. They did not look at whether water conservation or leak reduction would eliminate the need for additional capacity. They also did not look at any alternative locations, even though there were other places they could put the wells. They also already had substantial excess capacity, they just believed they needed more.

In another example, they wanted every one of their employees to have their own truck. They required every employee to report to a specific office on the southern side of the service area every morning and evening and thought they needed a work truck to accomplish this task. The service area was very large. They did not consider repairs to existing trucks they already owned, or having alternative locations that employees could report to so they would not have to drive so far every day. They also did not consider reimbursing personal vehicles. This cost was extremely large and approved by the board as it was stated that there were no alternatives. A bit of creative thinking could have greatly reduced their expenses. Ultimately, the utility was in such bad financial shape that it had to be taken over by another utility.

Projects on the short-term CIP plan need to be prioritized. This can be done based on risk, other criteria, such as the ability to meet the system's Level of Service goals, ability to complete the project in a timely manner, regulatory requirements, and other non-water asset activities. For example, a road project may be planned that would enable a pipe to be replaced at the same time.

Once projects are placed on the short-term CIP and prioritized, funding should be sought for the projects. This funding can be from the utility's own sources or outside funding sources and may be loan, grant, or a combination of both.

8.0 ACCESSING FUNDING FOR CAPITAL PROJECTS

Capital projects are large-scale, high-cost projects that require significant planning and design and are intended to improve, expand, or replace infrastructure. These projects typically require outside funding but can be paid for in part (or completely) by the utility's reserve funds. Outside funding generally comes from state or federal sources or bonds, although it may also come from a private bank, especially in the case of privately owned utilities. There are numerous sources of governmental funds, but the two most common and largest sources of funds (before 2025) include the State Revolving Funds (SRFs) and the U.S. Department of Agriculture Rural Development (RD) funds. The U.S. EPA more recently added a funding source for large dollar projects called the Water Infrastructure Finance and Innovation fund or WIFIA. The funds from federal sources are generally low-interest rate loans, but there may be a portion of grant funding available, although it may be termed "negative interest loan" or "principal forgiveness," rather than grant.

Another popular funding source is the Community Development Block Grant (CDBG) program that is grant rather than loan, but this program is highly competitive and relatively small dollar. Other agencies, such as the Bureau of Reclamation (western states only), Economic Development Administration, Army Corps of Engineers, and Indian Health Service (tribes only) provide funding for specific types of projects.

The number of state, federal, local and other agency programs for funding is quite extensive. The number of entities providing funds within a state ranges from five to 33 with a median of 12 entities per state. Because many of these entities have multiple options, the number of funding programs ranges from seven to 203 per state, with a median of 27 different programs. This large number of agencies and associated programs can be overwhelming to navigate, especially to determine eligibilities, timelines, requirements, and which funding source has the most favorable terms for the particular utility. (Funding Programs are listed by state on the Environmental Finance Center Network [EFCN] webpage at https://efcnetwork.org/.) Changes to infrastructure funding that started in 2025 may affect the number of programs available, the amount of funding, as well as the agencies that provide funding. There may also be reductions in the number of employees available to assist entities who are seeking funding.

The estimated dollar amount for total water infrastructure investment nationally far exceeds the available dollars for the various funding programs, so not every utility will be able to access the needed funding. Another issue is that not all utilities are in a position to take advantage of the available funding. The system may be resource constrained (both in terms of personnel and money), they may lack knowledge of funding sources and how to apply, or may lack the necessary technical, managerial, and/or financial capacity to take on a project, even one that is critically needed. Once a system is ready to proceed with a project, they may be able to find funding from one or more of the programs. However, it is not always the type of funding an entity may prefer, because most systems prefer a grant rather than a loan. For larger projects, or for projects requiring a match, it may be necessary to access funding from more than one source. One example is the WIFIA program that funds only 49% of the project costs, with the other 51% having to come from some other source.

Periodically, the federal investment has increased, such as the American Reinvestment and Recovery Act (ARRA) funding in 2009. This act included supplemental funding of $4 billion for the Clean Water SRF and $2 billion for the Drinking Water SRF, which was more than double the expenditures for 2008 for these funds (Ramseur & Humphreys, 2009). The funds included restrictions that the projects had to be "ready to proceed to construction within 12 months" (Ramseur & Humphreys, 2009). Additionally, a large portion of the funds were to be used for additional subsidies (i.e., grants) and a portion was dedicated to green infrastructure. Additional infrastructure spending was made available by Congress through the Infrastructure Investment and Jobs Act (IIJA) (commonly called the Bipartisan Infrastructure Law [BIL]) in 2021. BIL provided $50 billion to EPA to improve the nation's drinking water, wastewater, and stormwater infrastructure, making it the single largest investment in water from the federal government (EPA, 2026). Additionally, congress passed the American Rescue Plan Act (ARPA) in 2021, which is administered by the Department of Treasury, and included $350 billion in direct funding to all cities, towns, and villages in the United States. One of the eligible activities was "making necessary investments in water, sewer, and broadband infrastructure" (National League of Cities [NLC], 2026). These investments included funding for improving access to clean drinking water and supporting vital wastewater and stormwater infrastructure, projects on the Clean Water or Drinking Water State Revolving Funds (SRFs), lead remediation, stormwater infrastructure, and aid for private wells and septic systems (NLC, 2026). These funds must be spent by December 31, 2026.

While the terms of the funding may be very favorable—very low interest rates for 20 to 40 years depending on the specific program—the funds do not come without strings. There are requirements related to the application process (e.g., Preliminary Engineering Reports [PERs], Environmental Documentation) as well as the construction process (e.g., Davis-Bacon Act wage requirements and Buy American Build American provisions). These "federal cross-cutters" as they are sometimes called can complicate the funding process and add to the overall cost of a project. For these reasons, as well as others, some entities choose to use the bond market to fund their projects rather than state and federal programs.

Revenue bonds, in which the debt service is paid mainly from revenue generated by the water utility, account for the majority of bonds issued for water infrastructure. These bonds can be issued by many different entities, including state and local governments, authorities, commissions, special districts, or other units created by a legislative body that are intended to issue bonds to enable infrastructure construction. Revenue bonds are generally tax exempt and have interest rates higher than a general obligation bond that is backed by tax revenue (EPA, 2008). Some entities prefer the bond market to federal funding due to shorter payback periods and fewer regulations. However, the bonds require outside expertise, which can include legal advice and a bond counsel.

Many states have created their own funding programs from state revenues of one type or another to fund water infrastructure. These sources may include fewer regulations and strings than the federal sources. They may be more limited in dollars, but they can provide grant funding in some cases and may be more targeted to the specific needs in a state rather than federal funding which is established to address the needs across a much broader range of systems.

The process of accessing capital begins with a thorough understanding of the project itself. Any funder will need to know exactly what the project entails to understand whether the project fits within the scope of eligible activities and what type of funding might be applied. Customers and elected leaders need to be able to understand what will be constructed and how it will help the utility and its customers now and into the future. If the project cannot be clearly articulated, it is necessary to spend additional time determining the needs and the required solution before proceeding. If the project cannot be clearly articulated within the utility and to its customers, it is unlikely that a funding agency will be able to properly review the project.

The next step is to determine which funding sources may be able to fund the project as well as which funding sources the applicant (the water system) is eligible for. Generally, for any type of governmental water system, which can include municipal, authority, district, or similar types of organizational structures, there are multiple funding sources for which the system will be eligible. For homeowner association–type systems, which are more common in western states, the governing body is generally elected or appointed board members. These systems can be governmental in nature or private depending on the way they are created within the state structure. Some of these entities will be eligible for several funding sources, some very few. Private non-profit entities may be eligible for some governmental funding, depending on the state, while private for-profit entities are unlikely to be eligible for outside funding from governmental entities.

Once the list of eligible entities who can fund the project is determined, the next step is to choose which funding source or sources is the most advantageous to the water utility and its customers. This may be one source of funding or may include pairing up sources of funding to meet match requirements or to gain enough money to complete a project. Once the source(s) is selected, the application must be completed and submitted to the agency. Most agencies have staff that can assist with the application process and some non-profit entities can also provide assistance with this part of the process at no cost to the utility.

The application process is likely to include additional documentation that will need to be submitted at various points in the process, which can vary based on the entity from which the funds will be sought. They may include some or all of the following, as well as many other documents that are more specific to the funding source and are not listed here:

- Engineering documentation, such as a preliminary engineering report, project design, construction drawings, specifications
- Environmental documentation, such as an environmental assessment, categorical exclusion, or environmental impact statement
- Financial documentation, such as audited financial statements
- Asset management plan (drinking water) or fiscal sustainability plan (wastewater)
- Demographic information to support loan forgiveness or loan eligibility
- Rate information or rRate structure
- Existing financial debts/outstanding loans
- Governance structure

Additionally, some funding sources require a demonstration of adequate technical, managerial, and financial capacity/capability. Having a robust, implemented asset management program can enhance the utility's ability to demonstrate their capacity/capability and access the funding. If the funding is in the form of bonds, it may also reduce the cost of the borrowing if the system is able to get a higher bond rating from the bond rating agencies based on its advanced asset management program.

To complete the required documentation, the utility is likely to need outside professional assistance. This may include engineering, environmental science, legal, or financial experts. If these services are not already on retainer with the utility, it is best to procure the services through the use of a request for proposal (RFP) process. This process allows for an open process to select the professional who will best fit the community's needs. The most important part of the RFP process is to clearly scope the nature of the work with as little ambiguity and uncertainty as possible. Uncertainty creates risk for the professional which will result in a higher price. Furthermore, being unclear about the expectations is likely to lead to disagreements or confusion between the parties, which is not good for either side. The RFP should also include everything the utility will want the professional to do, especially if some of the items are not typical. For example, it is best if the engineering work includes a requirement for the engineer to provide asset inventory information in a format compatible with whatever inventory program the utility is using. The submission of asset inventory data should include all the attributes the system wants about each asset as well as replacement cost.

When assets are new it is the best, and easiest, time to collect data on them. However, if this task is not included in the RFP, it will not be included in the engineer's proposed scope of work and would constitute an added cost if it was added later. Additionally, the project should include information to update any mapping program with the locations of assets. It is important to specify that any information gathered by the engineer or other professional, such as mapping, locations, asset attributes is the property of the utility and not the professional. There have been cases in which the contractor created a map for the utility but the utility was only provided a paper copy and not an electronic copy. Then when the utility asks for the map electronic files, it may be told that it has to pay a fee to receive the files.

Once an entity closes on a loan or grant, there will be loan or grant management activities, including managing the construction bid process, the payment process, financial reporting, any required annual reporting, and construction oversight. As was described in the previous chapter, it is necessary to have construction inspectors who oversee the work and ensure that it meets the engineer's specifications. For smaller entities, these management activities can be difficult to complete and may require outside professional help or hiring additional staff. Larger entities need to balance the amount of work they do with the number of staff they have in-house to manage the projects.

If bonds are sought, instead of using a state or federal funding source, the process will be much different. This discussion is outside the scope of this book, but most entities who use the bond market are already familiar with the process.

The discussion in this section included funding sources available at the time of the book's publication. Federal and state funding sources change over time depending on the political climate and the overall economy at the state and national level as well as the perceived need to invest in water infrastructure. When an entity is ready to proceed with an infrastructure project, it is important to investigate the current set of funding sources. Similarly the amount of assistance from state and federal agencies to help water utilities with the funding process may vary over time as well as the entities who may provide free technical assistance.

9.0 CAP EX VS. OP EX

One element of world's best practice in asset management is the idea that a dollar is a dollar. If the utility intends to operate as efficiently as possible, it needs to implement the most effective options possible whether they are operational (op ex) in nature or capital (cap ex). It should not matter whether it is op ex or cap ex, just that it is the best thing to do. In many countries outside the United States, this thinking prevails.

Within the United States, it is a bit more difficult to consider that an op ex dollar is the same as a cap ex dollar because of the origins of the money. Op ex dollars almost always come from customers (user rates) while cap ex dollars can come from other sources. Cap ex can also be grant funding which is not possible for op ex dollars. Because some utilities want to keep user fees low, they may forgo proper operation and maintenance (defer maintenance) which leads to early deterioration of assets and the need for higher cap ex dollars. It is a way of transferring op ex to cap ex.

If a utility really wants to achieve all the benefits of asset management and the maximum efficiency, they need to see cap ex dollars the same as op ex dollars. It is also necessary for state and federal agencies to view the dollars this way as well. For example, providing grant funding or below market rate loans for utilities is a means of subsidizing—at the state or federal level—water utilities. This subsidization is considered reasonable and acceptable, but the idea of subsidizing operations is not, even if that option may be far cheaper over the long-term. Serious conversations should take place, as asset management practice matures, to determine how the United States can better align with the need to see cap ex and op ex dollars equivalently and to push for the most *overall* efficient operation.

10.0 PERSONNEL ISSUES RELATED TO LONG-TERM FUNDING

While operations personnel should be involved in budgeting and finance concerns, similarly, finance personnel should be involved in asset management planning and implementation. Nothing can be done without funding and financing, which makes financial personnel critical to successful asset management planning and implementation. Finance personnel should be included from the very beginning of the process and incorporated throughout. These finance personnel should see benefits from the asset management program and these benefits should be considered whenever the overall program is being assessed or benefits are reported to the public or elected leaders.

As discussed previously, operations personnel may ask for increased funding for operations, maintenance, supplies, equipment, vehicles, or capital projects. They may have great backup documentation to support the requested increases and believe that the increases should be funded. While asset management is one way to help build support for these requests, there is no guarantee the answer will be yes. What does it mean for asset management success if the answer is no? Utility personnel who experienced this situation, especially in the earlier days of asset management practice, have had responses to this occurrence something like "advancing our asset management failed" or "asset management failed." Often the morale declines, support for asset management wanes, and some wish to give it up entirely.

In actuality, it is not asset management that failed. Asset management provides the necessary data in a way that decision-makers can use to make the best decisions possible. The decision-makers have the ultimate responsibility to make the final decision. Often the decision-makers are balancing competing demands for the revenues as well as the overall community health and well-being and may not be able to make the decision that is most desired by operations and management personnel. The decision-makers may ask the operations staff to work within the budget constraints and hold off on projects. If negative consequences occur, the staff will need to work with the decision-makers to decide how best to respond. It is important for management to reassure operations personnel and keep the interest in asset management going even if the decisions are not as desired. The managers can look for other wins with asset management and use those to help continue support for asset management within the utility. Examples could include having a more planned day, fixing particularly problematic assets, improving a problem rather than just fixing it, or any other items that improve the operator's life at the facility.

11.0 INCREMENTAL INVESTMENT

Assets were installed in most utilities over a period of time with large investments being made at various points as the community expanded or regulatory changes required major changes to the treatment, collection, or distribution systems. Examples of regulatory changes that affected infrastructure investment are presented below.

- Requirement for Secondary Wastewater Treatment: Treatment plants had to add additional treatment units that have to be maintained and replaced periodically over time.
- Requirements for Tertiary Wastewater Treatment: Many communities are now required to remove additional contaminants, such as nitrogen and phosphorous, which requires advanced treatment facilities.
- Requirement for Surface Water Filtration: When surface water plants were required to add filtration, significant investments had to be made to add the additional treatment units.

- Elimination of Combined Sewer Overflows (CSOs): Initially many wastewater and stormwater systems were combined, especially in the East and Midwest. In these cases, both stormwater and sewage flowed in the same pipe and during high precipitation events, the combined flow would be directed to water bodies (rivers, streams, lakes, oceans). When regulations required the elimination of CSOs, systems had to invest considerable amounts of money to either build storm sewers, holding tanks, or larger treatment plants.

Installing large numbers of assets or large dollar projects at various points throughout the utility's existence means that the replacement of these assets is likely to occur in waves also. It may be difficult to obtain enough money to make these large investments. One way to dampen some of these peaks is to consider incremental investment in the assets. Rather than having very low capital investments for 20 years, then having to do a very large project, the utility may be able to invest an amount every year for the next 20 years to reduce the overall cost of the large project or eliminate it entirely.

For example, for a planning level estimate, consider how much pipe is currently in the ground for the sewer system, the drinking water system, and the stormwater system. Take that mileage of pipe and multiply it by an average cost per mile of pipe replacement for that community. Then consider what amount of money would need to be invested to replace that pipe entirely over a 100-year period. Then compare that amount of money to what is currently being invested, including the investment of money from any outside sources. Is the amount way below what it would cost to replace the pipe in 100 years? If so, there are very likely to be large spikes of investment needed or catastrophic failures are likely. If pipe tends to last longer or shorter than 100 years in the utility, a number higher or lower can be used. An example is provided in Table 6.2.

Consider that pipe is unlikely to last much beyond 150 years and some pipe will not even last 100 years. The costs are so high to replace significant

TABLE 6.2 Example of Calculation of Incremental Pipe Investment

System	Miles of Pipe	Cost per Mile (on Average)	Total Cost of Pipe	Investment Needed per Year to Replace Pipe Within a 100 Year Time Frame
Drinking Water	1200	$2.5 m	$3 b	$30 m
Wastewater	900	$4 m	$3.6 b	$36 m
Stormwater	760	$3 m	$2.28 b	$22.8 m

quantities of pipe that incremental investment should be considered to avoid hundred million or billion-dollar investments. It may take time to ramp up to these levels within a utility to prevent rate shock and to build up staff capabilities to administer this level of construction projects annually.

The same type of incremental investment may be a bit more difficult within treatment plants because technology changes, regulations change, and replacing only one portion of a treatment plant may be difficult. However, a robust asset management program should point out the opportunities that are available, such as replacing blowers, pumps, headworks, wells, and similar units.

The main point to remember is that all assets will eventually need to be replaced or removed from service and ignoring the problem will not make it go away. An investment will need to be made in new assets and that investment will either need to be made incrementally with a few larger spikes or made with many large spikes and periods of low investment in between.

12.0 REFERENCES

Environmental Financial Advisory Board. (2025, January 15). *Advancing water affordability nationwide: A framework for action.* https://www.epa.gov/system/files/documents/2025-01/efab-water-affordability-report.pdf

National League of Cities. (2026). *COVID-19 pandemic response and American Rescue Plan Act.* https://www.nlc.org/covid-19-pandemic-response/american-rescue-plan-act/arpa-local-relief-frequently-asked-questions/#eligible-uses

Ramseur, J. L., & Humphreys, E. H. (2020, July 24). *EPA water infrastructure funding in the American and Reinvestment Act of 2009.* Congressional Research Service. https://www.congress.gov/crs-product/R46464

U.S. Environmental Protection Agency. (2008, August). *Guidebook of financial tools: Paying for environmental systems.* https://nepis.epa.gov/Exe/ZyPDF.cgi/P100179D.PDF?Dockey=P100179D.PDF

U.S. Environmental Protection Agency. (2024). *Water affordability needs assessment: Report to Congress* (EPA 830-R-24-015). https://www.epa.gov/system/files/documents/2024-12/water-affordability-needs-assessment.pdf

U.S. Environmental Protection Agency. (2026). *Description of infrastructure investment and Jobs Act.* https://www.epa.gov/infrastructure/water-infrastructure-investments

Wichita State University & Southwest Environmental Finance Center. (2015). *KanCap water management manual: A guide for water boards.*

Chapter 7

Asset Management: It Works for Green as Well as Gray Assets

Asset management has traditionally been used to manage "gray" assets—those human-made elements of a water system, such as pipe, tanks, manholes, treatment units, meters, fire hydrants, valves, and pumps. However, there are often other elements of a water system that are natural or green assets, including land around a source of drinking water, forested land that protects a watershed, bioswales or rain gardens to collect and infiltrate stormwater, that are important to the overall functionality of the water system. These assets should be part of the asset management program to enable effective, holistic decision-making that incorporates all elements that are necessary to ensure regulatory compliance, proper operations, desired levels of service, and achieve the full range of triple bottom-line benefits.

Green assets are also commonly used in conjunction with gray assets in stormwater management systems—those systems designed for the conveyance, treatment, and discharge of stormwater. Sometimes, these types of assets are referred to as BMPs or best management practices. In this chapter, the terminology of "green infrastructure" will be used rather than BMP or any other common description of these types of assets/practices.

In general, green assets include anything that is meant to mimic natural processes. They can be divided into categories, such as the following:

- Natural Assets—When nature is being used to serve a function for the water system. These assets may receive some limited management, but generally are already in existence and left to perform their function as they are. These types of assets include: forests, streams, rivers, lakes, vegetated land around other assets, and natural wetlands.
- Green Assets (also called Enhanced)—Green assets (or enhanced) are human-made assets that use mostly nature-based materials (plants, grasses, trees, soils, rocks) that are designed and installed to perform a function similar to a natural process. These types of assets can include bioswales, tree trenches, retention or detention basins, infiltration basin, rain gardens, or constructed wetlands.
- Engineered Green—Engineered green assets are those that use human-made materials to mimic a natural process. For example, permeable pavement largely consists of human-made material and may initially appear to be a gray asset. The difference is impermeable pavement—the traditional gray asset—is not intended to infiltrate any stormwater, while permeable pavement is meant to allow stormwater to infiltrate into the subsurface below the pavement. Because these assets are attempting to mimic the natural process of infiltration into the ground, they are considered green rather than gray assets. A few other examples include green roofs, blue roofs, and rain barrels. In some circles, this type of asset is now being referred to as "blue assets" to distinguish the fact that these assets use human-made materials rather than just natural ones. However, this terminology is relatively new and not in wide circulation. Therefore, in this book, these types of assets will be referred to as "engineered green" rather than blue assets.

Collectively, all three of these categories will be referred to as "green assets" in this book and unless otherwise specified, any time "green assets" are referred to in this chapter, the terminology is meant to refer to all of these types of assets. Additionally, green assets are often collectively referred to as "green infrastructure," which should be considered in this book to be synonymous with green assets and the terms are used interchangeably here.

Green assets may be distributed throughout the community to achieve their purpose through collective action or they may be concentrated with the intent to provide a particular result in a given area. An example of distributed green assets would be a series of small rain gardens along roadways (Figures 7.1a and b). No one asset would be able to reduce a significant amount of stormwater, but collectively, they could reduce much more. This type of infrastructure creates more redundancy (if one small asset fails, the rest can still do their job) and may be the best way to fit green assets into a developed area, especially an urban one.

Green assets that are concentrated are generally over a large area and are intended to have a much bigger effect at each location. Examples of these types of assets would include permeable pavement over a large parking area or a large detention pond (Figures 7.2 and 7.3). In either case, the intent is to provide some primary benefit(s) related to water. The most common primary benefits include protection of drinking water sources, reduction of stormwater runoff, and improvement in stormwater runoff quality.

One question that may come to mind is why green assets should be incorporated into a traditional asset management program. There are several reasons for this approach. To make the most appropriate decisions about what solutions are best to address water, wastewater, and stormwater issues, a full array of options must be considered. These options may include green assets and without placing them within the asset management framework and thinking of them as assets, they may be forgotten during discussions of potential options. Another reason to include green assets is that once they are installed in a community, they should be managed as assets. They need to receive maintenance, repair, restoration (rehabilitation), and replacement similar to gray assets and if they are not included in an asset management system, the necessary work may be neglected or may not be included in the budget. Furthermore, although green assets serve a primary function related

FIGURE 7.1 Rain Garden Example

FIGURE 7.2 Large Detention Pond

FIGURE 7.3 Permeable Pavement

to water, as described above, they can have additional co-benefits that should be considered, such as social and environmental benefits. Including these assets into the asset management framework allows the comparison of gray asset solutions that probably serve only one function to green asset solutions that have primary and secondary functions.

Now that the "why" has been described, the next logical question might be how to include them. The best way to incorporate green assets into asset management is to incorporate them into a traditional gray asset management framework. Because there has been considerable work both in the United States and abroad on developing the asset management framework that is in practice now, including many companies who have developed software packages to address asset management, it would be extremely difficult to create a new process to replace the traditional approach. Furthermore, any system that has already implemented an asset management program would be reluctant to undergo a change to a new approach, not just because of the time and effort involved but also because there may be extensive investments of money and staff that may already be deeply imbedded in the current version of asset management.

For these reasons, the best approach is to imbed green assets into the traditional gray asset management framework even though it is not a perfect fit. The green assets should follow the same five core concept approach—level of service, current state of the assets, criticality, life cycle costing, and long-term funding—as for gray assets, although a bit of adjustment may be required.

While green assets do not fit perfectly within a traditional asset management framework, it is the best way to have overall management of both green and gray assets. A few of the complexities related to incorporating green assets into an asset management framework are shown on Table 7.1 below. A further explanation of these and other considerations is contained in Section 2.0, "Specific Considerations When Including Green Infrastructure into Asset Management."

Many, if not most, water systems probably have some type of green infrastructure somewhere in the system. It is important to consider including green assets in the traditional strategy. The most efficient and effective results will occur if all options (green and gray) are considered.

The most effective, efficient management of water infrastructure considers all options necessary—green and gray—to provide the environmental services and puts all options on equal footing in terms of decision-making. Each community should investigate the full range of options to address stormwater (both quality and quantity), source water protection, source water conveyance, and reduce combined sewer overflows and choose the combination of green and gray assets that provides the best overall system.

TABLE 7.1 Green Infrastructure Considerations

Asset Management Component	Issues/Concerns/Complexities
Level of Service	Setting goals around primary and secondary functions
	Setting goals that can be measured with easily obtained, readily accessible data
	Developing goals that balance customer expectations with capabilities
Current State of the Assets	How to define an asset
	Setting useful life of assets (especially natural that can have an indefinite lifespan)
	Determining asset condition
	Determining how asset condition can change over time (green assets may improve over time, whereas gray tend to degrade over time)
Criticality	Understanding failure modes of green assets
	An individual green asset is likely to have a lower risk than many gray assets
	Part of a green asset can fail without the entire function failing
	The difference between a single green asset failure and multiple asset failures
	Identifying the consequences of failure
Life-Cycle Costing	The costs of the various life stages of green infrastructure are difficult to estimate
	Natural assets do not have some of the costs associated with portions of the life cycle (may not have to purchase the asset and these assets will not be "replaced" or potentially disposed of)
	Green assets can be owned by different entities than manage them
	Maintenance may take a different skill set and need different staff and training than gray infrastructure maintenance
	There is less knowledge about maintenance needs of green assets
	Understanding a complete life cycle can be more challenging, especially with natural assets
Long-Term Funding	Funding is in its infancy with green assets
	Hard to develop a long-term funding plan
	Green assets can appreciate over time, whereas gray assets depreciate
	One problem with limited knowledge of funding is that green infrastructure is not being used to the extent it should be
	Green infrastructure options may be removed from consideration because of a belief that they "cost more" without actual proof that this is the case

Green assets can provide other benefits, in addition to the water-related benefits, which may add value (both monetary and non-monetary) to the community. Green assets should be managed effectively to ensure that proper maintenance is being conducted and that repairs or rehabilitation activities are conducted in a proactive manner.

Green asset management is in its infancy and will progress over time, similar to gray asset management. While green may be a bit behind gray asset management overall, there are many utilities that have not yet taken on gray asset management. Therefore, in these communities, it is possible to start right away with a combined green/gray approach. As the practice matures, more information will be available and green asset management will become more robust.

Not all public works directors or community engineers want to consider the secondary benefits of green infrastructure. Some of these individuals believe that their job is only to be concerned with the core functions of drinking water treatment and delivery, wastewater collection and treatment, and stormwater collection and conveyance. While it is understandable that their main focus is not aesthetics, climate change mitigation, health improvements, or the like, it is also important to remember that water infrastructure is the foundation of the community and anything that benefits the community is a benefit to water. If customers have a more positive image of the water infrastructure due to these secondary benefits, perhaps they are more willing to pay their current rates or support higher rates. Furthermore, a healthy community is beneficial to everyone, including public works, so consideration of the green infrastructure and its full range of benefits should be given greater weight.

Many communities have green infrastructure within their overall system but may not recognize it as assets to be managed. Others may be considering adding green infrastructure in the future. In either case, the best option is to incorporate green assets into the traditional asset management framework and manage all assets similarly.

Having a more expansive view of asset management that includes both green and gray opens additional possibilities and moves individuals away from the belief that only concrete and steel solutions will work. It will provide a more expansive view of the types of alternatives that can yield the desired results and the co-benefits that can be achieved. Some of the issues that may benefit from consideration of both green and gray infrastructure include the following:

- There are major concerns related to the decaying infrastructure (often referred to as "aging" infrastructure when decay is the real concern) and the cost that operating, maintaining, and replacing this infrastructure will entail. The affordability of maintaining a high standard of treatment for both water and wastewater is also a question, particularly in low-income communities and under-resourced communities. A more robust consideration of green options may help address these issues.

- Asset management and capital improvement plans help utilities plan for the realities of decaying infrastructure but do not generally incorporate natural assets (forested lands, buffer zones, green infrastructure). These assets can be an integral part of the treatment process given that the environment naturally absorbs and filters water via soils and plants. These assets can act as purifiers, barriers, and reservoirs.
- Pollution of our water systems continues to be a problem, particularly in large cities. As stormwater drains through gutters and storm sewers it is ultimately discharged into nearby waterways. These waters contain trash, bacteria, oil, heavy metals, and other pollutants. These floodwaters can also cause property, infrastructure, and habitat damage.
- There are many concerns, especially in the western United States, regarding water scarcity and security, including increasing temperatures that accelerate evaporation, reduction of groundwater and groundwater recharge, and reduced snowpack. All of these issues can affect the overall sustainability of communities. A greater consideration of natural systems and how they should be managed to help encourage water retention, aquifer recharge, and improved water quality can be part of sustainability planning.

These issues intersect in the concept of mixing green and gray infrastructure in a single asset management program. Often, the green and gray infrastructure are thought of and managed by different entities, governmental agencies, or different departments within a community. This approach has the negative effect of creating barriers to greater implementation of green or natural infrastructure. When capital projects are considered, the focus may be solely on the gray infrastructure, not the holistic picture that includes green options. If both the green and gray infrastructure were considered in the same asset management program, different decisions could be made. The green infrastructure could be valued appropriately for the benefits it provides. This combination of green and gray into a single asset management framework is not meant to value green over gray but rather to value green along with gray so that holistic effective, data-informed decisions can be made.

1.0 GREEN INFRASTRUCTURE TYPES

There are many types of green infrastructure that can be used for various purposes. The Southwest Environmental Finance Center (SW EFC) developed a green infrastructure database to provide information about green infrastructure and to expose interested individuals to a wider array of possibilities. The database can be accessed at the following address: https://swefcapps.unm.edu/gardb. Figure 7.4 is a screen shot of the database (as of 2025) (SW EFC, 2020).

The following types of information are included:

- Name of asset
- Asset type
- Water system classification
- Construction rank
    - Ease or difficulty of construction: 1—very easy, 2—easy, 3—medium, 4—hard, 5—very hard. Ratings are relative to the other assets in the table.
- O&M rank
    - Ease or difficulty of O&M: 1—very easy, 2—easy, 3—medium, 4—hard, 5—very hard. Ratings are relative to the other assets in the table.

FIGURE 7.4 Green Infrastructure Database

- How it works
- Operation and maintenance (O&M) required
    - Information related to the time commitment, how frequently maintenance is required
- Design considerations
- Costs
- Potential benefits (monetary as well as non-monetary)

The database is a searchable, web-based tool that allows utilities to search for green and natural infrastructure by a variety of characteristics including asset name, asset type, construction rank, and O&M difficulty. This tool serves as an introduction to the types of green infrastructure or natural systems that can be included in an Asset Management plan for water, wastewater, and stormwater and how they may be used to help address various regulatory and infrastructure requirements.

An example of the data that is included is provided here for one asset type, an infiltration planter. This data can be obtained by selecting the "view" button under details for the particular asset. (This example was current as of 2025. Updates to information occur periodically so the actual data may vary from what is provided below.)

- Name of Asset: Infiltration Planter
- Infrastructure Type: Enhanced (also called green assets in this book)
- Water system type that could potentially use the asset: Stormwater, Combined Sewer
- Construction Rank: 2 (easy)
- O&M Rank: 2 (easy)
- How it works: Infiltration planters are usually installed between the roadway and the right of way to manage stormwater runoff through infiltration, storage and evapotranspiration. They are commonly seen in cities along streets and sidewalks, and can vary in size, shape, and plant content. The infiltration zone in the planter is lower in elevation than the adjacent surfaces (i.e., sidewalks and roadway). These planters often have an overflow pipe connected to the existing combined sewer pipe for excess runoff. Stormwater planters are designed for small-scale runoff collection and filtration and should be located in areas where the water flow rate is not too strong. Hills and steep slopes usually cause problems for planters because runoff with high velocity will overrun the planter and not be able to infiltrate.

- O&M Required and Frequency: Water-efficient irrigation should be applied for the first two years after construction of the facility, particularly during the dry summer months, while plantings become established. Irrigation after these two years is at the discretion of the owner. Inspect plants and structural components periodically. Remove sediment and clear debris from inlet pipes and curb cuts to maintain proper drainage. Remove sediment and debris and replace mulch (this could be a frequent requirement if the facility design doesn't include pretreatment). Clean and repair inlets and outlets, embankments, and berm dams. Control erosion, replant as necessary, and remove weeds by hand.
- Design Considerations: Planter walls should be constructed of durable, impervious materials, but should not employ chemically treated wood that may leach chemicals into groundwater. Planters should incorporate trees and shrubs where feasible. An overflow should be installed to divert excess water during high-flow runoff events.
- Costs: Overall relative costs are expected to be low, with costs dependent upon construction materials and the size of the planter. Costs on a unit basis (cubic foot) are expected to be higher than other best management practices (BMPs). Costs vary depending on size and materials. For new development and redevelopment, infiltration planters are often less expensive than more conventional stormwater management facilities. Approximately $8/square foot, $400–$500/year for a 500-square-foot planter.
- Potential Benefits (monetary as well as non-monetary): Provides filtration of pollutants, as well as infiltration of runoff. Reduces flow rates and volumes. Suitable in areas with limited space. May be used as part of a traditional landscaping plan. Should reduce the amount of watering necessary to maintain vegetation. Infiltration planters typically require less piping than flow-through planters and a smaller facility size than traditional swales where native soils allow for infiltration.

Each of the assets included in the database can be included in an asset management framework, such as the one being described in this book.

2.0 SPECIFIC CONSIDERATIONS WHEN INCLUDING GREEN INFRASTRUCTURE INTO ASSET MANAGEMENT

The traditional approach to asset management was created for the needs of gray infrastructure, rather than green or green and gray in combination.

This fact means that putting green assets into traditional asset management will involve some consideration of the differences and how they affect each of the five core concepts. Some examples of concerns are shown below. This list is meant to be illustrative of the types of concerns but is not exhaustive.

2.1 Level of Service

- Setting goals that balance what the residents/customers want and ability to implement. For example, customers may not want any visible trash in the green infrastructure. It is most likely impossible to meet such a goal, especially when there are many green sites in the community or if there are frequent rain events that will drive trash into the sites. The goal may have to be to remove trash once per year or twice per year to keep trash levels from building up too high. However, when goals are not well-aligned with customer expectations, it is important to explain to customers the level of service that is possible with the given staffing and funding. If customers want higher levels of service, one option is to involve customers in some of the maintenance of the green assets, such as trash removal.
- While green and gray infrastructure may share many of the same overall goals in terms of basic water-related services, green infrastructure also has the potential to offer additional benefits, such as adding wildlife habitat, improving aesthetics, improving public health, increasing property values, and reducing the heat island effect. It is important to write goals around these secondary benefits, not just the primary benefits. Both customers and elected leaders should understand these secondary benefits and how well they are being achieved.
- As important as secondary benefits can be for the overall community and the utility customers, individuals who operate and manage water systems may not care as much (or at all) about the secondary benefits. Their primary job is the management of water, not the other factors, such as aesthetics and habitat. It is important to develop a culture within the water system and larger community of valuing these other aspects. It is important for community elected leaders to demonstrate the importance of the secondary benefits and to encourage water managers to incorporate them. If there is a particular project in which green infrastructure is more expensive than a gray alternative but provides greater benefits for the overall community, it is necessary for local elected leaders to voice their support for the higher costs and to provide the additional funding. In some cases, this additional funding may need to come from other sources, rather than the water, wastewater, or stormwater budget.

- Social aspects of green infrastructure are likely to be higher than gray infrastructure. Goals should be written around these aspects of the project.
- It is vital that goals be measurable. Some green infrastructure–related goals may be harder to measure and it can be difficult to gather the necessary data. However, the act of setting goals can lead to the development of practices and data gathering to aid in the process.
- Goals do not need to be overly complicated; keeping goals simple can aid in the development and use of the goals. The number of goals can also be kept manageable to make sure staff is not overwhelmed by them. It is more important to set a few important goals that can be measured, rather than to set lots of goals that end up ignored because it is too much work to keep up with them.
- Developing a few goals around green infrastructure and adding them to the overall list of goals can be a good place to start. Additional goals can be added over time.
- Developing goals around what are the most important aspects of the community's green infrastructure can also help in getting started. For example, if the goal is to reduce ponded water in a low-lying part of town, a goal can be written for that specific goal, such as: "No observable ponded water in the Elm Street area for any storm less than the 50-year rain event."
- If the community is particularly interested in green infrastructure related to resilience and/or climate change, goals should be written around these aspects.
- Consider whether the mission statement developed for the utility fully captures green infrastructure or whether it needs to be expanded in some way to incorporate its use.

The goals developed for asset management focus the utility on what is important and measuring and monitoring whether the goals have been met allow the utility to spend its limited resources on the most important aspects. Therefore, it is vital that if the community installs green infrastructure, at least some goals are written specifically around green.

2.2 Current State of the Assets

- One of the most difficult parts of adding green infrastructure to traditional asset management is deciding how to define green as an "asset." It is relatively easy to say a pump is an asset or a piece of pipe or a valve, but when a rain garden is installed, what is the asset? Is it the plants or the gravel/soil underneath? What about any underdrains? Are each of these sub-assets or is it one asset collectively? It is important for the utility to decide how it wants to categorize green assets. It can use whatever process makes the most sense but it should be reproducible by others, documented, and consistent. For example, if a rain garden is taken as one asset in one part of town, rather than multiple assets, it should be taken as one asset in another part of town. Procedures for determining what a green asset is for parents, children, and grandchildren should be developed and shared with all staff so that assets are properly and consistently categorized.

- In considering how to define an asset, it may also be important to consider what is serving the primary function (e.g., increasing infiltration) and what is serving the secondary function (e.g., plants to increase wildlife habitat). It may be helpful to have each of these elements designated separately as an asset.
- How the asset is managed can also have an effect on how it is defined as an asset. If it is managed as a single unit, it can be designated as one asset. If the components are managed separately, it should be broken down into multiple assets.
- The size of a green asset can help define how it is designated as an asset. A very small green asset may be one asset, while large assets are more likely to be broken into more than one asset.
- Natural assets will most likely need to be broken down into units, such as feet or miles of stream, or acres of forest land. The utility may only be interested in a small portion of the green asset (say the reach of river that runs through the city) or the portion of the forest that protects a water body. Only these portions of the natural assets need to be included in the asset inventory. However, if damage to the larger asset beyond the boundaries being considered, such as a wildfire, will affect the water system, it may be necessary to include a larger portion of the natural asset.
- How the overall asset is replaced can also play a role in how the asset is defined. It is important to break the overall asset into sub-assets if it is going to be replaced in separate pieces rather than as a whole.
- The types of attributes may be very different for green assets than gray. It is important to develop a list of what attributes are wanted for each type of green asset.
- Some green assets are intended to work together. For example, there may be several separate rain gardens along a street that are meant to work together to reduce stormwater runoff. The way assets are connected should be included in the asset inventory.

- The concept of useful life remaining is different for green versus gray assets. For natural assets, the useful life remaining may be indefinite. For other assets, the life spans of vegetation can be quite different than the life spans of some of the other components of the asset (e.g., the soil, underdrains, gabions, any concrete components). These differing life spans can be another reason to separate a green assets into various components.
- Assessing the condition of a green asset can be a bit tricky. In the case of green assets, the asset condition may improve over time, at least for a bit, due to vegetation growth, root maturity, and tree growth. It is important to consider what aspects an assessor should use to determine that asset's condition.
- Green assets can be owned and/or operated by other entities than the utility. It is important to note the ownership and/or operation entity within the asset inventory.

The most important aspect of the current state of the assets is to ensure that the green assets are included in the inventory. It may take a while to figure out the best way to include them and it may be necessary to change the process over time to fully accommodate the assets and ensure that the proper information is captured.

2.3 Criticality

- It is important to understand the failure modes of the types of green assets in your system. The failure categories are the same—mortality, level of service, capacity, and financial inefficiency—but how the actual asset fails may be harder to define. It is important to develop guidance for different types of green assets and how they can fail to meet their intended purpose. As an example, a forest natural asset could fail because an insect infestation occurred and killed a lot of trees, or a fire occurred, or a drought caused trees to die. Permeable pavement may fail if it becomes clogged, its underdrains fail, or vegetation grows in the cracks.
- Probability of Failure rankings need to consider whether an asset can continue to serve its intended purpose if portions of the assets fail. For example, a green roof is intended to have plants. If all the plants die, can the roof continue to serve its function? If a rain garden loses its plants, will infiltration still occur? If the asset can function, at least reasonably well, it should be given a lower probability of failure rating.
- Often, especially with very small green assets, the overall criticality (POF × COF) is low compared to gray assets. While this can

be discouraging to managers of green assets, since funding will go to those assets that are higher criticality, it is a good thing for the overall utility.

- Since green assets tend to be lower in terms of criticality, incorporating green assets into the overall system can help lower the overall risk of the system. This situation can be a driver for incorporating green assets.
- It is necessary to consider the effect of one individual asset failing versus several green assets failing when thinking of consequence of failure. If one green asset fails, the consequence may be low, but if a lot of green assets fail, it may be much more consequential. It is important to consider whether a failure in a particular type of asset is more likely to occur as an individual asset or a collective.

Green asset risk may generally be lower than gray asset risk, which is an overall positive thing for the utility. Developing an understanding of modes of failures and potential consequences is vital to determining risk.

2.4 Life-Cycle Costing

- The asset's life starts during planning, design, and construction and these early stages are extraordinarily important. Many failures of green assets can be tied to poor design or poor construction. If the planning process does not adequately include the interests of the stakeholders and community members, the green assets may not fit the community's culture and values that may affect their ability to ultimately meet the level of service goals.
- It is necessary to determine the costs of each phase of the asset's life cycle. However, there is limited information at this time regarding what these costs are. It will take a bit of time to adequately estimate these costs.
- Natural assets do not have some of the life-cycle costs. For example, the utility may not have to purchase a lake, or river, or forest; it will just be there. (In cases such as creating a lake just for the purpose of water supply, this dynamic will be different.) There may not be any "replacement" costs either as these assets are not going to be disposed of and replaced. However, natural assets can be rehabilitated (e.g., streambank restoration, forest restoration) or repaired (e.g., reparation after a fire).
- Generally gray assets are owned, managed, and operated by the same entity. However, it is entirely possible that different entities

own green assets than operate or maintain them. In some cases, a private entity may own an asset, but the utility maintains it or a different portion of the local government may own the asset, such as the Parks and Recreation Department or Transportation Department, but it may be maintained by the utility. Other assets may be owned, operated, and maintained by a separate entity with no involvement of the utility, such as private businesses. Because these assets are still intended to help the overall operation of the utility, it is important to consider them within the asset management strategy in some way, which may be just acknowledging that the asset, what type it is, and where it is located.

- There are equity issues/concerns with green infrastructure. It is important to ensure that green infrastructure is spread throughout the community so all can benefit and not just wealthy areas. The types of infrastructure may be different throughout the community to reflect the values and needs of the residents of the area.
- Maintenance may take a different skill set and need different staff and training than gray infrastructure maintenance. Overall, there is less knowledge about maintenance needs of green assets. There are some maintenance manuals that have been produced to help with maintenance and the utility may wish to develop its own to create a workforce that can perform the required maintenance. As green infrastructure evolves over time, knowledge around maintenance practices will increase and the workforce will be more available. If workforce issues are a major concern, tasks can be handled by an outside contractor, but the contractor must have knowledge and experience in the types of green infrastructure within the utility.
- Standard costs for green infrastructure are hard to develop because costs are very site specific.
- Understanding a complete life cycle and its associated costs can be more challenging, especially with natural assets.

Understanding green asset overall life-cycle costs will improve over time as more assets of this type are installed and used within communities throughout the United States. The improvements in cost will allow for more comparison of green and gray asset solutions. Some of the alternative analysis relies on cost estimates that may not be site specific and fully reflective of costs. The secondary benefits and other reasons to install green assets should come into the overall decision-making process. While not every facet of cost estimating is clearly understood for green assets, it is also true that not every facet of cost estimating is understood for gray assets and these costs are not always extremely detailed, particularly when considering the ongoing costs of operation and maintenance. It is important to not unnecessarily disadvantage green asset solutions simply because there is less information known about the life cycle costs.

2.5 Long-Term Funding

- The overall long-term funding process for green assets is in its infancy but will improve over time.
- Green assets may appreciate in value over time, whereas gray assets depreciate. This is particularly true if land values increase.
- There may be options to increase funding specifically for green assets through dedicated line items with water rates if customers desire green infrastructure and value its secondary benefits.
- It can be difficult to develop a long-term funding plan given that construction costs are very site specific, maintenance costs are not yet well known, operation costs are generally low to zero, costs of repair/rehabilitation/replacement are not well known and vary from place to place, and end of life is difficult to estimate. Over time, these difficulties will decrease.
- One problem with limited knowledge of funding is that green infrastructure is not being used to the extent it should be. Green infrastructure options may be removed from consideration because of a belief that it "costs more" without actual proof that it does over time.
- Funding for green assets needs to be included in the overall budget for the utility and it needs to be adequate to cover the costs of operating and maintaining the green assets.

Funding will be required to ensure that after green assets are installed they continue to meet their required level of service. Therefore, funding must not only include construction, but also maintenance, repair, and replacement.

There are particular considerations when it comes to long-term funding of natural assets. Water Research Project 4727, Asset Management Framework for Forested and Natural Assets, includes a discussion of the effects of accounting standards. This report describes the importance of the accounting structures governing the utility. The accounting structure has an effect on how to account for the money spent on natural assets as well as the types of funds (op ex vs. cap ex) that can be used to address natural assets. Therefore, "the accounting structure and associated standards can have a significant impact on developing the long-term funding plan for natural resources" (Raucher et al., 2020).

2.6 Conclusion

The most effective, efficient management of water infrastructure considers all options necessary—green and gray—to provide the environmental services and puts all options on equal footing in terms of decision-making. Each community should investigate the full range of options to address stormwater, source water protection, source water conveyance, and reduce combined sewer overflows and choose the combination of green and gray assets that provides the best overall system.

Green assets can provide other benefits in addition to the water-related benefits that can add value (both monetary and non-monetary) to the community. Green assets should be managed effectively to ensure that proper maintenance is being conducted and that repairs or rehabilitation activities are conducted in a proactive manner.

Green asset management is in its infancy and will progress over time, similar to gray asset management. While green may be a bit behind gray asset management overall, there are many utilities who have not yet taken on gray asset management. Therefore, in these communities, it is possible to start right away with a combined green/gray approach. As the practice matures, more information will be available and green asset management will become more robust.

Consideration of the secondary benefits of green infrastructure is important and funding the additional benefits, if initial costs of green infrastructure outweigh the costs of gray infrastructure, may be necessary from other non-utility sources. A healthy community is beneficial to everyone, including water utilities.

It is important to recognize that natural assets are rarely owned by the utility, even in cases where they may have a significant effect on the operations of the utility (such as water source, infiltration area, source water protection, or reduction in stormwater flow). Also, the unique characteristics of natural assets have an effect on how the risk is assessed, how levels of service are developed, and how investments are planned. These factors may require the utility to consider more of an iterative approach when addressing natural assets in an asset management strategy (Raucher, 2020).

Many communities have green infrastructure now but may not recognize these elements as assets to be managed. Others may be considering adding green infrastructure in the future. In either case, the best option is to incorporate green assets into the traditional asset management framework and manage all assets similarly.

3.0 GREEN INFRASTRUCTURE ASSET MANAGEMENT EXAMPLES

Included below are two examples of communities that have implemented green infrastructure asset management and some of the benefits they have received.

3.1 Tucson Water

Tucson Water, in Tucson, Arizona, formerly recognized stormwater as a water resource in 2023, not as consumptive use, but as an energy-free way to provide shade for the community. The Tucson Resilient Together Plan reflected that green stormwater infrastructure (GSI) would be an extreme heat mitigation strategy and a water resource/drought strategy. The program for green stormwater infrastructure is called "Storm to Shade or S2S" (Bushman, 2025).

The green stormwater infrastructure directs stormwater runoff from streets, parking lots, and rooftops into landscaped areas to infiltrate soils and support vegetation. This stormwater also helps local wildlife. Tucson Water uses a Tree Equity Score to choose where to place green infrastructure that includes: existing tree canopy, population density, surface temperature, race, income, employment, age, and health. In the future, as their asset management program matures, they hope to add criticality elements. The intent is to bring all areas of the city to the same standards over time (Bushman, 2025).

As part of the asset management program for green infrastructure, in 2022 and 2023, Tucson Water inventoried over 700 GSI assets and included geolocation, asset and sub-asset type attributes (size, presence, capacity, quantity). The asset inventory information is included in both GSI and their enterprise asset management software. They also established a maintenance program for GSI in 2021 and 2022 and published a GSI Maintenance Manual. Using the manual, they trained six contractors with specialized skills and have a Memorandum of Understanding between their department and the Streets and Parks departments to coordinate between different asset management systems. More than 1100 maintenance visits were performed as of January 2025. As part of maintenance of GSI assets, Tucson puts a medallion on each asset that shows the name of the program (Storm to Shade) and includes a contact phone number. They also place a placard out in the street or sidewalk to show customers what their funding is being used for. The language on the placard is "Your Green Stormwater Fees at Work" (Bushman, 2025).

The Storm to Shade (S2S) Program joined the Tucson Water Digital Transformation process to include GSI assets in the same asset management system. The green asset management system includes a planner/scheduler that can also schedule gray asset work along with the green infrastructure and a system administrator who configures new assets, manages new users, runs report, and provides troubleshooting. The addition of green assets into the gray asset management framework was the first in the history of Tucson Water and created some challenges. Storm to Shade was the largest user of contracted labor and involved building living assets into the contractor training and checklists that were designed for gray assets. The Tucson Storm to Shade program used the Integrated Asset Management Framework developed by the SW EFC and the Green Infrastructure Leadership Exchange, Asset Management Peer Learning Circle as resources to develop their green infrastructure asset management program (Bushman, 2025).

The Storm to Shade program developed level of service goals for the program. The goals were intended to ensure safety and included: infiltrating standing water; pruning vegetation for site visibility, mobility access and hazards; preserving structural integrity; and clearing trash, invasive plant species, and debris. These level of service goals enabled S2S to rehabilitate some green assets that were installed by other departments but overrun with invasive species and otherwise not functioning properly (Bushman, 2025).

As part of the current state of the assets portion, S2S determined that each of its seven green asset types would represent a parent asset. Examples of parent assets include: traffic circle, right-of-way basin, chicane/curb extension, and inlets/outlets. The four-times-a-year work orders are written to the parent asset. Each asset has children or sub assets, but the children assets are not configured separately. For example, if there are five trees, the inventory indicates five trees but does not call out each tree separately due to the difficulty of naming, numbering, or identifying living assets individually. Follow-up and ad hoc work orders can be written on children assets. Examples of children assets include: inlets/outlets, curb cuts, curb core, sediment trap, infiltration basin, and planting shelf. Data is collected on any new green assets using a collection application and an asset ID is autogenerated for the asset. Once the data is collected, an asset update request is sent to the asset inventory program manager and the GSI system administrator configures the asset data in the system, including the date it was commissioned, the route for maintenance, and the preventative maintenance schedule (Bushman, 2025).

Contractors were provided training on the use of the asset management system for maintenance. During training the contractors were given real-time assistance to make sure they all understood the system. They were provided mobile applications to allow the use of the system in the field. There is also a computer application that can be used in the office. As of 2025, there were 810 GSI assets in the system that were managed by Storm to Shade.

Each priority asset receives a maintenance visit four times per year, but the contractors can also put in a work order request for additional maintenance if necessary based on the results of the field visit. Additionally, if follow-up work needs to be performed, such as replacing a broken asset or installing a missing asset, a follow-up work order can be written (Bushman, 2025).

Successes thus far for the Tucson green asset management program include:

- Automating the steps required to deploy individuals to perform maintenance work on GSI, saving considerable amounts of time.
- The asset management system holds a lot of data that can be used to track performance and forecast long-term funding needs. They have used a financial forecasting tool to forecast out the next 5 years of costs as capital projects move to maintenance work. Data from the asset management tool has been used for this purpose, especially costs related to specific dollars to perform specific tasks.
- Ability to help small contractors grow their business base. (Bushman, 2025)

Challenges of the green asset management program include:

- Configuring multiple green/living sub assets in GSI is difficult.
- The field application of the asset management program was cumbersome for contractors to use.
- Training entry level contractor staff on the asset management technology and the GSI skills required a lot of training time and investment. (Bushman, 2025)

Refinement and next steps for the GSI asset management program include:

- Shortening field checklist to follow-up work and data collection on invasive species and tree data.
- There is now data to develop criticality and life-cycle costing portions of green asset management and that work will be upcoming.
- The city of Tucson did an urban tree inventory of the whole city, not just GSI-related trees, and they would like to integrate this inventory into the asset management software. (Bushman, 2025)

In summary, "An effective asset management system will support S2S in ensuring that equitably distributed GSI assets FUNCTION and PROVIDE the multitude of benefits that Tucsonans need to buffer our experience of climate change—increased drought and heat, and more variable precipitation and flooding" (Bushman, 2025).

3.2 DC Water

The District of Columbia Water and Sewer Authority, commonly called DC Water, was under a consent decree to address its combined sewer overflows (CSOs). It initially intended to build three large holding tunnels to increase capacity during high rainfall events. However, DC Water determined that the costs of construction, as well as the uncertainties associated with future rainfall events, warranted a reconsideration of the approach. Ultimately, DC Water determined that it should install green infrastructure within the Rock Creek Park corridor to replace one of the tunnels (U.S. Environmental Protection Agency [EPA], 2025).

Installing green infrastructure projects in the corridor, rather than the holding tank, resulted in triple bottom line benefits that started in 2017 instead of starting in 2024 (the time frame expected with the tunnel). The benefits are anticipated to be related to increased property values, neighborhood beautification, heat island effect reductions, natural habitat enhancement, improved public spaces, and economic development through local green jobs (DC Water, 2025). The revised plan, including green infrastructure, "is expected to reduce CSO releases by 96%, based on current precipitation levels" (U.S. EPA, 2025).

To manage the green infrastructure being installed, "DC Water is developing a robust asset management framework to plan proactively for green infrastructure maintenance. The utility began its green infrastructure asset management program in 2013. Now it is managing existing assets with the three-fold goal of ensuring performance, public and maintenance crew safety, and aesthetics" (Water Environment Federation, 2015). DC Water is planning to use asset management software to manage the green assets, including maintenance of these assets (Water Environment Federation, 2015).

4.0 REFERENCES

Bushman, B. (2025, January 23). *Nature as infrastructure—City of Tucson GSI Enterprise asset management system.* Southwest Environmental Finance Center Green Infrastructure Webinar Series: Asset Management for Green Infrastructure. https://www.youtube.com/watch?v=5pcfVhawrCM, https://efcnetwork.org/event/webinar-quarterly-green-infrastructure-webinar-series-asset-management-for-green-infrastructure/

DC Water. (2025). *DC Water's green infrastructure*. https://www.dcwater.com/resources/environment/cleanrivers/projects/green-infrastructure#:~:text=GI%20can%20increase%20property%20values,Jobs

Raucher, R. (2020). *Asset management framework for forested and natural assets* (Project 4727). Water Research Foundation.

Southwest Environmental Finance Center. (2020). *Green asset resources database*. https://swefcapps.unm.edu/ggdb/

U.S. Environmental Protection Agency. (2025). *DC utilizes green infrastructure to manage stormwater*. https://19january2021snapshot.epa.gov/arc-x/dc-utilizes-green-infrastructure-manage-stormwater_.html

Water Environment Federation. (2015, November 4). Evolving green infrastructure through asset management. *Stormwater Report*. https://stormwater.wef.org/2015/11/evolving-green-infrastructure-asset-management/

5.0 FURTHER READING

Bernalillo County. (2022, July). *Bernalillo County green stormwater infrastructure/low impact development strategies*. https://www.bernco.gov/public-works/wp-content/uploads/sites/76/2022/08/02-BernCo-GIS_LID-Appendix-A.pdf

Charles River Watershed Association. (2008, September). *Low impact best management practice (BMP) information sheet: Stormwater planter*. https://nacto.org/wp-content/uploads/stormwater_planter_crwa.pdf

City of Lancaster. (2019, February). *Green it! Lancaster* (Appendix B). https://www.cityoflancasterpa.gov/wp-content/uploads/2023/09/GREENIT_LANCASTER_FINAL_withAppendices_022519.pdf

City of Lincoln Transportation and Utilities. (2020). *Stormwater post construction best management practices guidance manual*. https://www.lincoln.ne.gov/files/sharedassets/public/v/1/ltu/utilities/watershed-management/best-management-practices/bmp-guidelines.pdf

City of Portland, Oregon. (n.d.). *Infiltration planters*. https://www.portland.gov/ppd/infrastructure/managing-rain-your-property/planters

Clean Water Services. (n.d.). *Low impact development approaches handbook*. pp. 30–33. https://cleanwaterservices.org/development/dnc/lida/

Green Building Alliance. (n.d.). *Stormwater planters*. https://www.gba.org/resources/green-building-methods/site-solutions/stormwater-planters/

Green Infrastructure Ontario Coalition, Green Infrastructure Leadership Exchange, & Southwest Environmental Finance Center. (2021, December). *Green stormwater infrastructure asset management resources toolkit*. https://giexchange.org/wp-content/uploads/2021/12/GSI-AM-Resources-Toolkit-Final-Dec-17.pdf

Chapter 8

Asset Management Implementation

The benefits of an asset management program cannot be realized if the elements are not implemented. The water system should observe behavior change as a result of its actions. Additionally, the asset management program must be sustained over time. Activities and actions must be put in place to ensure that the program continues over the long term. The program cannot be associated solely with a few staff members or just an "asset manager." It must be driven deep within the organization so changes in staffing, resignations, retirements, and the like do not cause the program to fail or to be abandoned.

1.0 THE ASSET MANAGEMENT PLAN: WHAT IT IS AND WHAT IT ISN'T

An asset management plan (AMP) is often seen as the "destination"—the place you are trying to get to with asset management implementation, but that is not the case. It is important to remember the concept that asset management is a journey, not a destination, and as such, the AMP is a waypoint along that journey. The AMP addresses the gaps between where you are now and where you want to be.

Essentially, the AMP describes where the utility is today, where it wants to go, and the actions it is going to take to get there. As an example, the utility may not have all of its assets included in the asset inventory, or the asset locations/attributes may not be accurate or complete, at the current time. The utility specifies what type of inventory exists today, what is included, and what is missing. The utility wants the inventory to be more complete in the future so it lays out the actions that will be taken in order to achieve that end point. As an example, these types of actions might include:

- Every time an asset is dug up in the field for repair or inspection, attribute and location information must be verified or collected as part of the work order. No work order can be completed without including this verification that the inventory is correct or without providing new information if the inventory is currently inaccurate. Additionally, as part of the work order, locational information must be verified to ensure that the asset is in the correct location on the map. A box on the work order must be checked if any changes need to be made to the inventory to alert the proper staff members of the need to make a correction.
- All visible assets will be inspected to gather attribute and location information on a schedule of one fourth of the assets inspected each year until completed. Thus, the inventory should be completed in approximately 4 years.
- For all attributes that could not be determined in the field, personnel will be assigned to try to find the information through searches of old files (paper or electronic), asking knowledgeable personnel (either current staff or former staff), asking manufacturers, or other appropriate methods. The amount of effort to put into this activity should be commensurate with the value of the information. If the information is highly useful, then it makes sense to spend a fair amount of time on the research. If the information is moderately useful, less effort should be expended. If the information is of limited value, only a small amount of time should be invested in trying to find the information or the information may be deemed unnecessary.
- The Asset Manager and the GIS Specialist positions are tasked with revising the inventory based on the information provided in the work orders or the data searches.

- An assessment will be made annually regarding how much data is still missing as well as what data types are missing. The goal will be to reduce the number of missing data points by 20% from the year before.

The actions described above are those that will move the utility from *what is* (a deficient inventory) to *what should be* (a much more robust and complete inventory). Being as specific as possible about the actions to be taken, how they will be taken, and over what time frame allows everyone to be clear about the plan going forward.

In terms of what information to include in an asset management plan, the contents may be slightly different for just about any organization that undertakes an initiative to improve its asset management practice. However, there are several themes to include that can be modified or expanded as necessary to meet individual needs. There can also be other sections added to address other interests important to the system. An example of an AMP outline is included in Appendix A. Highlighted below are some of the key themes of the AMP.

1. Executive Summary: The Executive Summary should be written to provide an overview for elected leaders, the general public, and other stakeholders who may wish to understand the overall process but do not need the details provided by the plan.
2. General Statement of the Commitment of the Organization to Asset Management: It is important that the organization's commitment to asset management be recognized and understood. If that commitment is in the form of a resolution or other official document, that language should be included along with any pertinent details (date resolution passed, vote tally). This section can also include any information regarding the resources (personnel and financial), that are being dedicated to the effort by the leadership of the organization. If the organization chooses to develop an AM Policy or Risk Management Policy in addition to or instead of a resolution, those documents should be included also.
3. Demand—Current and Future: Describes the demands that the utility currently operates under and the changes that are expected in the future. If changes are expected, it will describe how the system will address these changes.
4. Level of Service: The Mission Statement for the organization should be included in this section along with the level of service goals of the organization related to the asset management program. For each

goal, the target level and the method of measuring the goal should be described along with the schedule upon which the goals should be reported and to whom. There should be a recognition of which goals are internal versus external and how the external goals will be reported.

5. Current State of the Assets: This section should describe the current status of the asset inventory/asset registry as well as the actions that will be taken to improve the asset inventory/registry over time. This section can describe the attributes included in the inventory for each class of assets along with which attribute fields are currently populated for at least some assets in the class. There are two asset attributes that should be described more fully—condition and useful life remaining. The status of asset mapping should also be described along with any actions that will be taken to improve the status of mapping information. This section can also include where the information is stored/where it can be found. For larger systems or those that are a bit more advanced in their asset management practice, information about the utility's net asset value over time can be included. This information may be used to determine if the utility's bottom-line value is increasing, decreasing, or staying the same and whether assets are failing to maintain their value. This latter information can be a valuable consideration when elected leaders are trying to decide about infrastructure investments.
6. Criticality: The methods by which assets are assessed for criticality should be described. This includes a description of how probability of failure (POF) was assessed (1 to 5, 1 to 10, other?); how consequence of failure (COF) was assessed (1 to 5, 1 to 10, other?); whether POF or COF were weighted differently or the same (i.e., it is possible to weight POF or COF more heavily if that is desired by the utility); whether redundancy was included as one of the factors considered in POF and/or COF or whether it was included as a separate factor; what the overall risk profile looks like currently; what the desired risk profile is; and the overall risk tolerance of the organization.
7. Life Cycle Costing: This section may vary a bit from organization to organization. It is the most complicated part of asset management and covers many elements. It may build over time as utilities get used to asset management. It should cover all parts of the asset's life cycle, including planning, design, construction, operation, maintenance, repair/rehabilitation, replacement, and disposal, and discuss the utility's practices, policies, and procedures in each phase and the

potential interventions that will be undertaken in each. The ways in which the interventions may extend asset life or reduce asset failures should also be considered, including how the utility will determine the benefits of the actions over time. The section can also include how the utility will consider the full life cycle costs when evaluating project alternatives.

8. Long-Term Funding: This section includes information related to funding the full cost of operation of the facilities and how that will be accomplished. This section can discuss budgeting, general sources of funding, user rates, capital improvements planning, and similar financial aspects. The section can also include information on the utility's capital improvement planning processes (both short and long term) and describe the business case evaluation process the utility uses for high-dollar capital projects (if one is used).
9. Staff/Staffing Considerations: This section includes information about the staff and staffing aspects of the organization, including current staffing levels, future staffing needs, succession planning, and related items. Staff related costs represent a large portion of a utility's operating budget and the system will not be able to meet the level of service goals, including regulatory requirements, if there are insufficient staff to operate the facility. Additionally, staff are required to plan for asset repair/rehabilitation/replacement and oversee the planning, design, and construction process. The human side of the facility is vitally important and should be given considerable attention. The staff should be given support to understand how they fit into the overall asset management program for the utility and to assist them in change management as the facility moves forward with its asset management program.

How the asset management plan fits into the overall planning process may vary from organization to organization. A small to medium-sized organization, or an organization just beginning its asset management journey, may have only an asset management plan that covers the entire organization and all of the assets and activities included within that plan. Larger organizations or those with more experience in asset management may have different levels of documents, including an Asset Management Strategy Document, a Strategic Asset Management Plan, and asset-specific Asset Management Plans that describe implementation on an asset class basis. In this format, the Asset Management Strategy answers the question of "how" you will achieve your goals with asset management. The Strategic

Asset Management Plan answers the question of "what" you will do. The individual asset-specific AMPs will describe how the utility will move from "what is" to "what is desired" for each asset class. For example, there may be AMPs for storage tanks, pump stations, wells, hydrants, pipes, and valves.

This multi-document approach can be too complicated for an organization that is new to asset management and this level of complexity is not necessary when one is just starting out. Instead, the asset management plan can be created that is essentially a combination of the three types of documents described above.

The KISS principle, which in this case means Keep it Simple and Sustainable, is very important when it comes to asset management plans. It is not necessary, or desirable, to have a voluminous document that no one is going to read or understand. It is also not necessary to put a lot of asset management background and theory into the document. A short description can be included, but that is all that is needed. There are plenty of sources for this information (including this book!) so no need to repeat that material. Links can be included to resources, if desired, to show the reader where they can find this background information. Instead, the document should be focused on utility information and how the utility will be conducting its activities. The document does not need to be any longer than necessary to convey the information and detailed materials should be moved to an appendix to keep the overall plan short and to the point.

The plan should also be written generally enough so that it is not out of date the minute it is released. It is best if the plan can be in effect for a period of years (5 to 10) before it needs revision. In this regard, general information about the overall condition of assets (what percentage of assets are in excellent, good, average, fair, and poor condition) is better than putting the specific asset conditions. These specific conditions can be provided elsewhere so that they can be updated at any time, but at least annually. The AMP should be considered a strategic document that shows general information rather than specific data at this moment in time so that it can last longer. The plan should be reviewed periodically and updated as necessary, which can be every 5 to 10 years or so. It will probably be too time consuming and overwhelming to review and revise the plan annually and it should not be necessary. Likewise, having a review period longer than 10 years is also unwise. Many things are likely to be different after a decade and the plan will need to reflect those changes. Additionally, if the plan is implemented during the 10-year period, the maturity in asset management should be much improved. This improvement needs to be reflected as well. Furthermore, technology changes rapidly and within a decade, there are likely to be significant changes that need to be reflected in the plan.

2.0 ASSET MANAGEMENT STAFFING

It is important to consider the staffing level of the utility overall and the staff that will be specifically involved in implementing asset management. If the utility is significantly understaffed, either through a high number of vacancies or because the budget does not support enough positions to fully operate the facility, it will be more difficult to implement asset management. As part of overall asset management implementation, the utility should consider addressing staffing shortages.

Asset management involves behavior change, and as such, it will modify the way business is done and those changes will affect staff in various ways. The utility needs to consider how it will assist staff members in adapting to the changes. This may include assigning staff within the organization to assist with changing management, using another department within the city/county/larger organization the utility is a part of (e.g., human resources), or hiring an outside entity for a period of time to assist with the initial adjustments.

Utilities will benefit from having a "champion" for asset management. This champion is someone who fully supports the goals and objectives of asset management and wants to see the utility implement this program. The "champion" may be anywhere in the organization. They should be supported by upper management to allow them to push for the necessary changes in practice.

It is desirable to have some employees within the staff dedicated to asset management activities. For medium-sized utilities, a three-person group is ideal—one person to run the overall asset management program (the Asset Manager), the second person to handle geographic information system (GIS) and all associated activities (GIS Analyst), and the third person to handle the asset management software (Asset Software Manager). For larger organization, the same structure can be valuable but there may be more than one person in each group. For example, there may be an Asset Manager as well as assistant asset managers, there may be several people involved in the GIS mapping and several helping to implement the asset management software. In very small organizations, there is likely only one person available to handle asset management and it may be necessary to contract out GIS mapping and have a lot of technical support from an asset management software company. In very small utilities, the job of asset manager is likely just one of the tasks for a staff member who has many other obligations.

The Asset Manager is intended to oversee the implementation of entire program and assist with change management. This person should be high enough within the organization to be able to direct employees to provide data or assist in the program or otherwise direct activities. This person should also be able to help with analysis of data and provide ideas of what types of things to analyze. The position can also be used to review how well goals are being met, review the Asset Management Plan, oversee the work of the GIS Analyst and AM Software Manager, meet with any employee who is struggling with asset management implementation or who has suggestions of improvements that could be made with asset management, and should be a communicator within the organization and to the elected leaders.

The GIS analyst should be in charge of all mapping activities, including entering data into the map and revising any data that was initially incorrect. The analyst can also create maps or views for others within the organization who want to visualize data. The analyst can also help with providing data in handheld devices for employees to use in the field (i.e., phones or tablets) to help in finding assets and accessing useful resources. The analyst can work with staff to ensure that data is user friendly and to design a process to gather any data on necessary map revisions.

The AM Software Manager is intended to help implement whatever software program the utility chooses to use, including helping employees access the software, ensure it is user friendly, set up dashboards, act as a liaison to the software company to get questions answered, perform any required updates or backups, and correct any erroneous data in the inventory. The Software Manager should also be able to help with technology transition. Some employees may not be used to interacting with computers, tablets, or phones (at least for operation and maintenance purposes) and may need help learning how to perform basic functions related to the asset management software, including data collection; data usage; work order development, completion, and closing; and asset inspections.

These positions may be outside hires or they may be internal ones chosen from within the utility based on skills, capabilities, and interests. There are pros and cons to both approaches and either can work well. Sometimes, the best approach is to find people with the correct mindset, such as any "champions" within the organization, and provide additional training needed to allow them to fill any of the required roles in asset management. Once the employees are chosen, it is important to set up a succession plan so that if any of them leave their positions or leave the organization, they are able to be replaced without adverse effect on the asset management implementation.

Because employees are so important to a successful implementation of asset management practices, it is important to think through appropriate staffing in general and staffing for asset management in particular as part of the implementation process. It is also important to provide training and support for all staff, including the asset management-related personnel. Training in asset management practice should be provided to all new employees at the time of hire or shortly thereafter. The training can be provided by an outside entity or by internal staff. The Asset Manager could be a likely choice to conduct this training.

3.0 GETTING STARTED

The best thing about asset management is that it can start anywhere and at any time. It is never too late or too soon to begin the asset management journey. The past does not need to hold anyone back. The system is where it is and that cannot be changed—the system has the assets it has, they are where they are, they are made out of whatever material—so it is best to just use the past as a learning opportunity. It is not something to lament or think, "We cannot do asset management because we are starting from such a bad place." Chances are the situation is not as bad as it first seems and there is always a way to make the situation better.

Although the journey can start at any point (remember asset management is a circle), there are a couple of logical starting places. Each of these is described below. Where you start may be determined by what feels the most comfortable, what seems to be realistic and doable, what is likely to result in a quick win or benefit, and what fits your particular experience with the water utility.

Level of Service Mission Statement and Goals: An excellent starting point is developing a mission statement for the organization. What is the utility trying to achieve? What holds it together as an organization? What are the most important values? What do the customers expect from the utility and what is the most important? The mission statement process is described earlier in the book and should be relatively easy to create.

Following the development of the mission statement, select a few goals, perhaps 10 to 12, and just start using them to manage the utility. Measure progress towards meeting the goals and determine what changes would need to be made if the goals are not met. Make sure the entire organization understands the goals and their role in meeting them.

An Asset Map: If the organization does not have a good, useful map of its assets, creating one is an important first step. As discussed in Chapter 3, there are many tools that can help in making a map. Some tools are free while others are low cost.

An Inventory of a Specific Class of Assets: While a full asset inventory may be an overwhelming starting point, choosing a specific class of assets to inventory first can help get the program started. The first class should be one that is easy to find and observe, such as meters, hydrants, pumps, or tanks, so that a relatively complete inventory can be created. Once one class is done, another can be chosen. Eventually, they can all be put into the inventory.

An Inventory of a Specific Part of the System: Similarly, rather than doing a specific class, a specific geographic part of the utility—such as one pump station and all of its affiliated assets or a portion of a water or wastewater treatment plant—can be chosen. Once one portion of the system is done, additional geographic areas can be done.

4.0 IDENTIFYING GAPS AND PRIORITIZING ACTIVITIES

This section may be more advanced than a smaller utility may want to consider, especially at the beginning of the process. However, those more advanced or medium to larger utilities may wish to have a process to identify their gaps and prioritize asset management activities. This section provides an option for utilities to consider if they want to use a systematic approach.

When advancing asset management thinking, there will be activities that will need to be completed to create a robust asset management program or sustain the program over time. There may be items identified that stand in the way of accomplishing all tasks or receiving the desired benefits. The gaps can be determined through a gap assessment or by holding meetings with staff to discuss those items that make asset management difficult. There may be many gaps identified when asset management is in its early stages. As asset management implementation continues, there may be fewer gaps. Some examples of potential activities are shown below for illustration purposes.

Examples of activities that may be identified:

A. Establish an asset management committee, along with logistics of how the committee will do its work (how and when it will meet, for how long, who should be included, etc.).

B. Create a process for assessing risk of vertical assets.

C. Create a process for assessing risk of horizontal assets.

D. Create standard operating procedures (SOPs) for what data should be collected in the field on assets and how it should be collected.

E. Develop a process for ensuring errors found in the field are corrected in the asset inventory and/or asset map.

F. Develop a data quality control/quality assurance (QA/QC) process.

G. Develop a process to determine if maintenance activities had the desired effect (e.g., longer asset life, reduced down-time, improved performance after activities are completed).

H. Incorporate Green Assets into the Asset Management program.
I. Develop a process for revising condition ratings after completing work on assets.
J. Develop SOPs for completing condition ratings on a routine basis (determine what time interval constitutes "routine").

Even if a relatively short list is identified, such as the one shown above, it will not be possible to fill in every gap or address every item all at once. These activities need to be prioritized so that they can be addressed in a proactive, planned manner along with a schedule that is reasonable (not so fast as to overwhelm the participants or too slowly to prevent progress).

One method of prioritizing activities is using an urgency/importance/level of effort matrix. Urgency refers to how quickly something must be dealt with and how it affects the ability to continue implementation of asset management. Importance refers to how much that item affects the overall creation and or sustainability of the asset management program. Level of effort refers to the time and resources it will take to complete the task. Each of these elements can be ranked according to high, medium, and low based on the criteria shown on Table 8.1.

Each activity identified can be ranked according to the three criteria. It should be noted that high for importance and urgency each rank 3 points while high for level of effort ranks only 1 point. This rating scale is because the items that should rise to the highest priority are those that are: highly urgent, highly important, but take few resources. These items would rank high urgency (3 points), high importance (3 points), and low level of effort (3 points) for a total of 9 points. The prioritization points for various levels of prioritization are shown on Table 8.2.

Table 8.3 shows an example of prioritizing each of the activities that were discussed earlier in this section based on priority points for each criterion. The letter related to the activity is shown in the table rather than the long descriptions, which matches with the list presented previously in this section.

In practice, the rankings should be completed by a cross-sectional team of personnel from within the system. There are no absolute right or wrong answers; the rankings should reflect the best estimate of urgency, importance, and level of effort based on the group's core beliefs. Once the rankings are completed, the group can review the overall rankings to make sure that there are no additional changes that should be made. Additionally, the cross-functional team will need to "ground truth" the prioritization plan with utility leadership such that there is agreement regarding priorities, resource allocations to implement the plan, and consistency with other utility strategic efforts. Then the activities can be sorted from highest score to lowest to show the overall priority rankings. As an example of this process, the activities from the example are put into priority order in Table 8.4.

Once the activities are prioritized, they can be completed in the order in which they rank, similar to what is shown on the table above. For items that have the same score, they can either be further prioritized to choose which one goes first, second, third, or they can be done simultaneously.

TABLE 8.1 Prioritization Criteria

Element	Description	High	Medium	Low
Urgency	Speed with which the activity must be dealt with	Must be completed within 1 year to continue to make progress with the asset management program (3 points)	Needs to be completed after 1 year and before 3 years (2 points)	Can be completed after 3 years (1 point)
Importance	How much the item affects the overall creation or sustainability of the asset management program	Foundational activity/element to the overall asset management program; mission critical (3 points)	Necessary to sustainability of the overall asset management program; major, but not critical activity to the asset management program (2 points)	Will assist in asset management program implementation or sustainability over the long term; minor contributor to asset management program (1 point)
Level of Effort	Time and resources required to complete activity	Significant amount of time and/or money to complete; complex task; multiple staff; takes more than a year to complete (1 point)	Moderate level of effort; may involve few or multiple staff, could take up to a year to complete; could take other outside resources (2 points)	Minimal effort required; simple task requiring only a few staff and a short amount of time (a few months) (3 points)

TABLE 8.2 Prioritization Points

Prioritization	Urgency Score	Importance Score	Level of Effort Score
Highest: 9 pts	High: 3 pts	High: 3 pts	Low: 3 pts
Moderately High (7 or 8 pts)	High or Med (3 or 2)	High or Medium (3 or 2)	Low or Medium (3pts or 2 pts)
Moderate (5 or 6 pts)	Medium or Low (2 pts or 1 pt)	Medium or Low (2 pts or 1 pt)	Low or Medium (2 pts)
Low (3 or 4 pts)	Low or Med (1 or 2 pts)	Low or Med (1 or 2 pts)	High or Med (1 or 2 pts)

TABLE 8.3 Prioritization Points Example

Activity	Urgency Score (3 = High, 1 = Low)	Importance Score (3 = High, 1 = Low)	Level of Effort Score (1 = High, 3 = Low)	Total Score	Comments
A	3	3	3	9	Establishing an asset management committee is fundamental to all other activities the system will undertake. It is both highly urgent and important. It will take relatively few resources or time to establish.
B	3	3	2	8	Risk is a fundamental part of asset management and should be done at the beginning of an asset management program. It may be complex and involve several staff members.
C	3	3	2	8	Same as B
D	3	3	3	9	It is important to have a process for what data will be collected on the assets out in the field. This process can be used for initial data collection on visible assets and a process for assessing non-visible assets when they are uncovered during repair work. While the act of collecting the data may take time and resources, this activity is just to develop the process and should not take a lot of resources.

(continues)

TABLE 8.3 Prioritization Points Example (*Continued*)

Activity	Urgency Score (3 = High, 1 = Low)	Importance Score (3 = High, 1 = Low)	Level of Effort Score (1 = High, 3 = Low)	Total Score	Comments
E	2	2	3	7	While it is important to create a process to fix data when errors are found in the field, it may be less urgent and important than other tasks. It should take very little in the way of resources.
F	2	3	3	8	It is important to have good quality data in the asset inventory system and it should take relatively little resources to create the QA/QC process.
G	1	2	1	4	For the sustainability of the asset management program, it is important to have a sense of whether maintenance activities are having the desired effect. This cannot be done quickly as it will take extensive time to see the effects of the maintenance activities. This is a complex task requiring multiple staff considerable time.
H	1	2	1	4	If the system has green or natural assets, it is important for them to be added to the asset management program over time. This probably cannot be done immediately and will be complex and require multiple staff.
I	1	2	2	5	It is important, but less urgent, to develop a process to revise condition ratings after work has been done on the assets, particularly for pipe assets that are often forgotten. It will take moderate resources to complete.
J	1	2	3	6	In addition to revising condition ratings after work is done, it is necessary to have a process to revise condition ratings periodically over time. This is not urgent, but it is important, and should take limited resources.

TABLE 8.4 Activities Ranked by Priorities

Letter	Activity	Score	Priority Rank
A	Establish an asset management committee, along with logistics of how the committee will do its work (how and when it will meet, for how long, who should be included, etc.).	9	1
D	Create standard operating procedures (SOPs) for what data should be collected in the field and assets and how.	9	1
B	Create a process for assessing risk of vertical assets.	8	3
C	Create a process for assessing risk of horizontal assets.	8	3
F	Develop a data quality control/quality assurance (QA/QC) process.	8	3
E	Develop a process for ensuring errors found in the field are corrected in the asset inventory.	7	6
J	Develop SOP for completing condition ratings on a routine basis (determine what time interval constitutes "routine").	6	7
I	Develop process for revising condition ratings after completing work on assets.	5	8
G	Develop a process to determine if maintenance activities had the desired effect (e.g., longer asset life, reduced down-time, improved performance).	4	9
H	Incorporate Green Assets into the Asset Management program.	4	9

In this example, the first activities are to form an asset management committee and develop a protocol for gathering data in the field. The asset management team could be used to develop the protocol so it should be done first followed by the creation of the SOP for data collection. The SOP can be the first activity for the newly formed asset management committee.

In an actual system, unlike this simple example, it is highly likely that the list of activities will be much longer, perhaps numbering 30 to 150 activities. It is important when ranking to avoid ranking everything high importance and high urgency. It is not possible to do everything first, so the activities need to be separated into high, medium, and low for each category.

As each activity is complete, it is important to acknowledge the accomplishment and share any positive outcomes or benefits received. The successful completion of these activities should result in an improved work environment for employees and build momentum for continued implementation.

It is possible that other activities will be identified that need to be added to the list. At the time of their discovery, they can be ranked according to the prioritization system and be added to the list at the appropriate location.

5.0 RIGHT-SIZING ASSET MANAGEMENT

Asset management, at its core, is a way of thinking, not a particular software, tool, action, or procedure. It can be done in a relatively simple way or an extremely robust manner. The simplest installations may use spreadsheets and simple data collection. The most robust installations may use sophisticated software, machine learning, digital twins, complex data analysis, and numerous condition assessment techniques. The most important aspect is not the complexity, but rather the ability to match the level of practice or level of maturity to the needs of the organization. This is referred to as "right-sizing asset management," meaning that the asset management program meets the requirements.

Each portion of asset management—the five core components—can have a different level of desired maturity and even within the components the maturity can vary for different activities. The basic maturity levels are presented below.

- *Aware:* At this maturity level, the organization is marginally aware of the concept or activity but has done no activity to advance asset management in this area.
- *Basic:* The organization is conducting some activities in this area at a minimal or beginner level. There is limited to no analysis of information and little strategic thinking around the activities beyond the basics.
- *Core:* Organization is doing activities that meet a foundational level of asset management, meaning the activity can support additional analysis of data and use of other aspects of asset management.
- *Intermediate:* Activities undertaken go beyond the foundational level but do not yet meet the level of utility best practices. Complex analysis can take place.
- *Advanced:* The activity undertaken is at the level of U.S. utility best practices or world best practices. High level of sophistication in all aspects, including staffing, data collection, analysis, and use of technology.

The organization can assess its current maturity level (where it is today) as well as its goals for where it ultimately wants to end up. It does not, and for many organizations probably should not, set "advanced" as the goal for each activity. It is important to assess where the organization really needs to be and what it is able to achieve when setting the future maturity level goal. Not every system needs to be at the level of U.S. or world's best practice. This is particularly true for smaller communities with more limited staff and resources. A core level of practice may be entirely appropriate for these entities. The main idea is to think about the maturity desired and match the practice to that level, including filling in the gaps when practice is less than what is desired.

A simple example of assessing maturity, including current practice and future goals is shown in Table 8.5.

The example above shows that the organization has a range of maturity levels across asset management based on each core component. It is possible to break the component further into categories and assess the current maturity, as well as the desired future maturity, for each of the subcomponents.

The organization should develop a plan regarding how to move from its current level of maturity to its future maturity level goal but should be mindful of not over-expending resources to go beyond the stated goal. For example, the current state of the assets goal is only "core," so very sophisticated asset management software is unnecessary. Instead, a basic computerized maintenance management system or asset management software that includes the basics without a lot of additional analytical tools will most likely be sufficient.

Sticking with the example of asset management software, there is a tendency to assume that the more sophisticated the program the better. This attitude can lead an organization to buy more software than they are capable of using that could lead to frustration and over expenditures. If the software is "right-sized" for the organization, it will meet the needs of the current situation and future goal for maturity level without requiring additional resources (time and money) to implement.

Several utilities have purchased highly sophisticated asset management software programs that have capabilities far beyond the level of the personnel to implement. The use of these capabilities requires data, QA/QC processes, and technical know-how. This situation leaves many elements of the program unused. To avoid purchasing the wrong product for the situation, it can be helpful to identify the maturity level goal before embarking on the software selection process.

The idea of "right-sizing" asset management practice is relatively new. Earlier on, most people believed the goal was always to move towards best

TABLE 8.5 Example of Using Maturity Levels Within an Organization

Core Component	Current Maturity Level	Future Goal for Maturity	Gap Between Current and Future Maturity
General AM	Basic	Intermediate	The personnel have a working understanding of asset management terminology and the basics of an overall asset management program. Ultimately, the organization would like to move to an intermediate level with a high level of sophistication in its understanding of asset management and would like to imbed asset management into all levels of the organization.
Level of Service	Aware	Core	The organization is aware that level of service goals should be set but has not begun that process. Ultimately, the organization would like to set basic goals and use them to assess its ability to meet customer expectations.
Current State of the Assets	Core	Core	The current practice includes a basic asset inventory and asset data. The data is deemed sufficient to perform all the remaining aspects and no change from current maturity is planned.
Criticality	Basic	Intermediate	The current practice is at the basic level with just a simple understanding of criticality and how it affects the organization. The organization believes criticality is at the heart of implementing asset management and wants to achieve a high level of maturity in this area.
Life-Cycle Costing	Basic	Intermediate	Life cycle costing is at the very basic stages and no analysis has been completed using asset data. This area will greatly benefit the organization and an intermediate level of maturity is desired. It is anticipated that an investment will need to be made into technologies to support this effort.
Long-Term Funding	Core	Intermediate	Funding maturity meets the needs of the current organization at the level of core, but will not meet the needs of the organization as it moves to more sophisticated asset management. It will need to be advanced in maturity as the asset management practice progresses.
Implementation/ Asset Management Plan	Aware	Core	The organization does not have an asset management plan and has only implemented the beginnings of asset management. While a plan is desired to aid in implementation, a core level is considered adequate for the future.

practice and initially, to world's best practice. This goal was unrealistic for many entities and greatly increased the resources that would be required to implement asset management. As the idea of selecting the most appropriate level of asset management maturity gains traction, resources can better match asset management requirements and implementation levels may increase across the water sector.

However, because advanced asset management practices are still relatively new for many water systems, the maturity level remains below what many systems require. A 2020 American Water Works Association (AWWA) survey of water and wastewater systems on the current state of asset management practice included a question regarding whether the respondent believed the asset management maturity level matched what was required/desired. Only 11% of respondents indicated that the level of practice was what was needed for the utility. The respondents were also asked to state which component of asset management had the greatest gap between current practice and maturity goal and responses were fairly evenly split between the five components with a slightly higher number indicating life cycle costing and long-term funding (which were combined on the survey) was the biggest gap and a slightly lower number choosing level of service (AWWA, n.d.).

6.0 MEASURING PROGRESS IN ASSET MANAGEMENT

There are several methods of assessing progress in asset management. One simple method was developed by the Southwest Environmental Finance Center (EFC) and is called the Asset Management IQ (AM IQ) tool. This tool is freely available to anyone wishing to use it and consists of 30 questions that assess practice in five core components as well as general asset management. There are 5 questions in each section and each question has answers scored from 0 to 5 points. The "0" answer is for little to no practice in that particular area while the 5 point answer is for a high level of practice. The tool can be accessed at the following address: https://swefc.unm.edu/home/amiq/main.php (SW EFC, 2011).

The idea is to use the IQ at the beginning of the asset management journey to assess the baseline of asset management and then annually (or other periodic basis) to show progress. The tool allows the user to save or print out the summary scores as well as the individual answers to all the questions to have a record of the evaluation. A printout from the AM IQ tool is shown in Figure 8.1. The printout shows that the total score for Waterville is 78 out of a possible 150. The weakest areas are Section 1 (General Asset Management) and Section 4 (Assets Critical to Sustained Performance.) The system may want to work on these areas first as they show weakness or

NAME: William Water

UTILITY: Waterville

February 8, 2022

9	Your score for section 1: General Asset Management
18	Your score for section 2: Current State Of The Assets
14	Your score for section 3: Level Of Service
6	Your score for section 4: Assets Critical To Sustained Performance
15	Your score for section 5: Life Cycle Cost
16	Your score for section 6: Financing
78	**Your Cumulative Score**

FIGURE 8.1 Progress Based on Sections

they may wish to work on the highest area of practice, Section 2 (Level of Service), to develop confidence before addressing their weaknesses.

A system can check progress over time to see how the asset management program is progressing on individual components or on asset management as a whole. An example of progress is shown below. The first graph, Figure 8.2, shows progress over time for each of the 6 sections. The second graph, Figure 8.3, shows progress over time for the overall program.

There are other tools that can be used to assess progress in asset management as well. Some may be free of charge while others may require membership in an organization or some fee to use. In any case, it is a good idea to use some methodology to measure progress in asset management over time.

7.0 ASSET MANAGEMENT AND ITS TIES TO REGULATION

One of the primary duties of a water system operator or manager is to ensure that all regulations are consistently met; protecting public health and the environment is of paramount importance. Customers want to trust that their drinking water is safe and healthy to drink, that their wastewater is

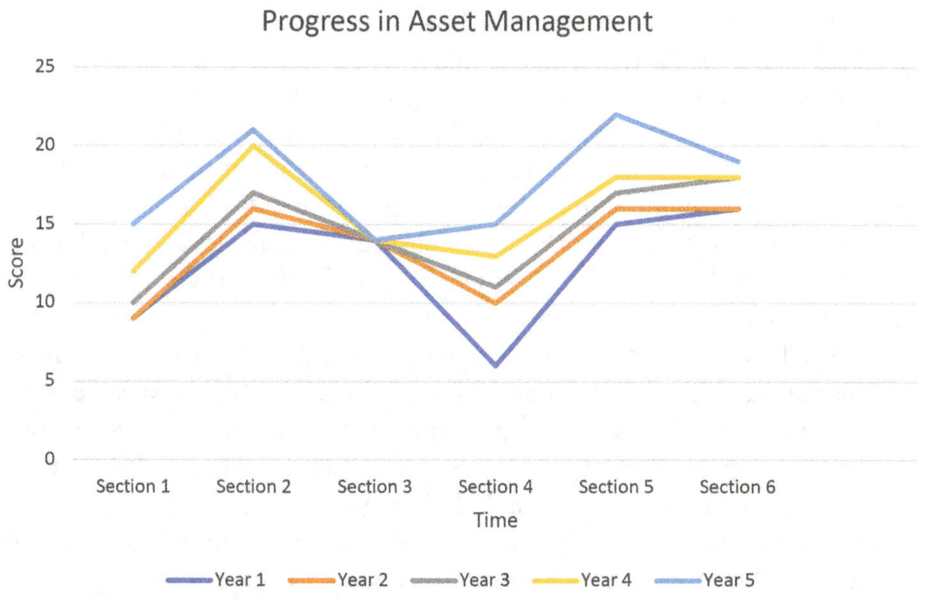

FIGURE 8.2 Progress Over Time for Each Section

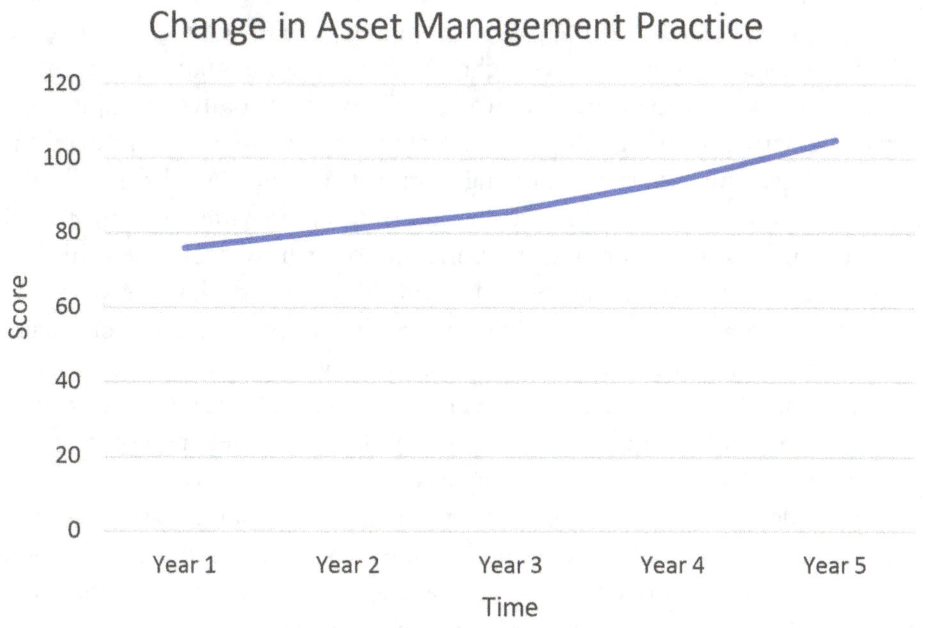

FIGURE 8.3 Change in Overall Asset Management Practice Over Time

being properly treated to ensure a healthy environment and that stormwater is being handled to best protect the community and the environment. Asset management is intended to help utilities do their basic functions well, which in turn ensures that utilities meet their regulatory requirements. The basics include: protecting public health, protecting the environment, providing customer service, maintaining assets in acceptable condition, and managing risk. All of these elements work to help ensure compliance with regulations and safe and healthy water, properly treated and disposed wastewater, and adequately handled stormwater.

The individual portions of asset management have a role in ensuring compliance is achieved and maintained over time. Some of the specific ways each part of asset management helps ensure compliance are discussed below.

Level of Service: Part of Level of Service is setting goals and one of the goals that every utility should include is a goal to meet regulatory compliance. In this way, the utility will be tracking compliance regularly and reporting it to decision-makers and elected leaders and discussing reasons why compliance has not been met if that indeed happens. The utility also has the ability to set a compliance goal that is more stringent than the regulations require.

Furthermore, setting the external goal of meeting compliance can lead to setting internal goals to help meet compliance. This connection can be referred to as developing "line of sight" between internal goals important to utility staff that can be used daily (or routinely if not daily) to adjust performance with external goals that are reported out infrequently to external stakeholders. This external reporting demonstrates how well the utility is meeting its commitments. Seeing the connections provides operation and maintenance staff a clear path to understanding how their work affects the performance of the entire organization. Making sure that the system is tracking its ability to meet these internal goals will provide another means of assuring continued compliance. Level of service goals around compliance can also be reported to customers on an annual basis to ensure that customers are able to hold the utility accountable for meeting compliance.

Current State of the Assets: The asset inventory as well as the asset map provide the system with information regarding what assets the system has available to meet compliance and where they are located. Furthermore, when the asset inventory is being created, some of the attributes that will be assessed are the condition of the assets and the useful life remaining. For assets that are critical to meeting compliance, it is important to know when the condition is heading in the direction of inadequate or the asset is reaching the end of its useful life. This information is important because failure of these assets may cause the utility to fail to meet its regulatory requirements.

Understanding condition can enable a utility to take appropriate actions to maintain the condition above a minimum level. The actions may include repair, rehabilitation, or replacement of some parts. For assets that are reaching the end of their useful life, it may be unacceptable to allow the asset to fail, as this action may cause the system to violate permit limits or other regulations. In these cases, the system will have to take action to replace the asset before failure. This replacement must be included in the capital replacement plan.

Criticality: It is likely that assets that rank high for criticality have the potential to have an effect on the system's ability to meet compliance. Systems should carefully evaluate the assets that are highly critical to determine which ones have the potential to result in a violation of permit limits or health standards or any other state or federal regulation. Any asset that does have that potential should be repaired or rehabilitated as soon as possible to keep it working well enough to continue meeting standards until it can be replaced (unless the repair/rehabilitation will fix the issue).

Life Cycle Costing: An asset's life begins during planning and design. This is the time that assets can be evaluated for their ability to consistently meet compliance. The methodology proposed can also be examined for the amount and type of operation and maintenance that would be required to meet applicable regulations and permit limits. It would be in the system's best interest to ensure that the system is capable of handling the required tasks before settling on the alternative. A different alternative that perhaps required less staff involvement or less sophisticated operation might be a better choice for a small system that may have trouble obtaining a higher-level operator.

Construction is another opportunity to affect the system's ability to comply. Assets need to be installed correctly and as intended to function properly over the long term. Having construction inspectors on-site, including regular visits from system operational staff, can help ensure that the assets start out well-positioned to serve their function and meet compliance.

The proper operation of assets is essential to meeting compliance. Having a good operation and maintenance plan that discusses how the system is meant to operate routinely and during emergencies can help prevent or limit any regulatory failures. Furthermore, while asset management is primarily focused on assets, it is possible to use asset management to gather operational data and assess trends over time. This data can be very helpful to relate operational data to conditions at the utility and compliance. What trends happened before the compliance issue, for example? Were these conditions present at other times when regulatory compliance was a concern? Once trends are examined, necessary changes can be assessed and initiated.

If compliance issues occur, work orders can be examined to see if failures in particular assets led to the compliance problem. They can also be examined to see how failed assets before a regulatory upset were repaired. Were any of the repairs inadequate to meet required performance? Did that lead to the regulatory failure?

Asset replacement is a big part of life cycle costing. Choosing when to replace assets can have a big effect on consistent regulatory compliance. It is necessary to have a capital improvement plan that specifies when assets are to be replaced over the short and long term. The short-term detailed plan should include any asset that is necessary to continue regulatory compliance. The long-term plan should include additional assets that will need to be replaced over the next 50 to 100 years to indicate to the elected leaders/decision-makers the fact that incrementally investing in and replacing assets is necessary to ensure the system operates as it should.

Proper disposal of obsolete or decommissioned assets is another place to ensure compliance. Improperly abandoning pipe might lead to a cross connection at a later time or homes being connected to the wrong pipe. Improperly abandoning a well might lead to contamination of the wells that are still in service.

Long-Term Funding: Asset management helps to make the case for incremental investment in the facility so that assets can be replaced on a more routine basis before significant failures occur. These incremental investments can help ensure that regulatory compliance is maintained.

Adequate revenues are also an important component to the proper operation of the facility, including funding staff, supplies (such as chemicals), repair, rehabilitation, and replacement. Most of these revenues will come from user rates and asset management is one aspect of determining what those rates need to be.

Summary: While asset management is not the only way a system meets and sustains compliance, it is one important component to the overall process. Having a robust, implemented, thoroughly imbedded asset management program will aid with long-term, consistent compliance.

8.0 ASSET MANAGEMENT AND ITS TIES TO FUNDING

While day-to-day expenses are paid by the system's rate revenue, capital asset replacement can use either the system's own funding or outside funding. Asset management can help justify increases in either of these types of funding. Asset management identifies the lowest life cycle cost of operation that ensures customers' expectations are met. Having sound data and information to justify the need for and magnitude of rate increases can build confidence in elected leaders to vote for the necessary increases. It can also blunt criticism of the increases from customers since they will be informed of where the money is going.

A small utility had traditionally been afraid to raise its rates. The governing body did not want to vote for them even though the system needed additional funding. After the utility undertook an asset management program in which it was able to clearly demonstrate what funding was needed to keep the system properly functioning as well as why it was needed, the utility was able to have a thorough conversation with its customers about those needs. The utility determined what the rates needed to be, the governing body approved the increase, and not a single customer complained about the increase. This was a big change from the past.

Bond rating agencies consider a utility's asset management practice when determining the organization's bond rating. Improving asset management practice may lead to improvements in the organization's bond rating, which can lower the borrowing costs if bonds are used to pay for the project.

9.0 ASSET MANAGEMENT AND RESILIENCY

The basic definition of resiliency is the capacity to recover quickly from difficulties. This definition can be revised slightly to say it is the ability to *withstand* and *recover* from outside stressors. Outside stressors can be any event or activity that has the potential to disrupt normal operations. These stressors can include natural events, such as fires, floods, hurricanes, tornadoes, ice, snow, or heavy rains; public health events, such as pandemics; human-made events such as a cyber security issues, terrorism, vandalism, or disgruntled employees taking some action; utility outages, such as a long-term electrical outage; climate change–related issues, such as drought, extreme heat or cold, or excess rain; or even financial events, such as the sudden loss of revenue. It is almost certain that every system will, at one time or another, face at least one, if not more, of these stressors. The main concern is how well the utility plans for events before they happen and responds to the stressor during and after the event.

If a water system has developed and implemented an asset management program, it has many of the tools needed to better withstand and recover from the stressor. Some specific ways asset management can help are described below, but there are many other examples that could be provided.

- Asset management includes mapping assets. Knowing the location (especially latitude and longitude) of assets can be extremely helpful if the event is a flood and debris ends up collecting on top of assets or the event is a tornado and landmarks that are normally visible are no longer useful in finding asset locations. Being able to find assets quickly can enable the utility to be up and running sooner than if a search has to be done to find important assets, such as wells.

- In addition to locations of assets, it is useful to have knowledge of all of the assets in the system to be able to know what assets are available during the emergency, what condition they are in as well as their criticality. For example, if a pump needs to be moved from one location to another to enable a failed pump station to operate, it is necessary to know how many of the similar pumps there are, where they are located, what condition they are in, and whether or not they are able to be moved (e.g., are they redundant or primary). Following a disaster, having a list of assets can be beneficial if it is necessary to deal with state or federal agencies after the disaster, such as the Federal Emergency Management Agency (FEMA).
- Part of asset management is determining which spare parts are necessary to have on hand for emergency repairs or replacements. It is particularly important to have spare parts available for assets that are highly critical and in those cases in which it is difficult to get parts or for which there is a long lead time. It is equally important to know where those parts are located within the facility, since having spare parts will not do any good if they cannot be located and quickly. A well-organized, appropriately stocked spare parts inventory will enable the utility to be more resilient to outside stressors that cause damage to assets.
- Asset management involves understanding the condition of assets and how many assets fall within different condition ratings, from excellent to poor. If the facility has a large number of fair and poor condition assets, it is less likely to handle any kind of physical disruption (e.g., fire, flood, heavy rain, tornado) than if the assets are in better condition. Understanding asset conditions and reviewing the overall condition of assets can help ensure they are kept in reasonable condition at all times.
- Asset management is about doing the basics well. It is about ensuring the functions of water delivery, wastewater collection and treatment, and stormwater management meet the basic customer requirements. Meeting the basics will allow utilities to better respond to any outside stressor that might occur.
- Having an operation and maintenance (O&M) plan, which is part of life cycle costing will ensure that operators and managers know what actions to take, when, and how, if they need to fill in during

an emergency in an area of the facility that is less familiar to them. For example, during a pandemic, operators may get sick or have to quarantine and someone else will need to fill in. The more information available to help the people who need to assist, the more resilient the utility. In addition, the O&M plan should contain information about emergency operations to assist in quickly addressing the issue at hand. The emergency procedures should be easy to locate, without the need of the main operator, in case the main operator is not able to go to the facility.

- Setting Level of Service (LOS) goals can help make a utility more resilient as well. The utility can set goals for how it intends to respond to emergency situations to aid in communications with the customers and to indicate the way it intends to respond.
- Some outside stressors can cause financial impacts to the utility, such as a tornado that destroys a section of the town or a pandemic that puts people out of work for a while. In these cases, revenue may be reduced or altered for a period of time. For example, the COVID-19 pandemic changed usage patterns from commercial to residential. In systems in which residential rates are lower than commercial rates, the change in usage has the potential to result in a reduction in revenue. Asset management can build resiliency by having a long-term financing strategy that includes having reserve accounts, cash on hand for emergencies, and that routinely covers full costs of operation. The more the system covers its costs on a routine basis, the better the system will withstand financial pressures during emergencies.

While asset management may not be thought of as a program to build resilience in utilities, it definitely is. It is one more incentive for systems to advance asset management practice and if an outside stressor event occurs, the system can benefit greatly from having a robust program in place.

10.0 TOOLS AND RESOURCES AVAILABLE FOR ASSET MANAGEMENT

There are many resources available to assist with completing an asset management plan. One resource is the Asset Management Development and Implementation Tool that was created as a companion to this book. Other resources can be found on the Southwest (SW) EFC website under resources, by selecting "asset management" as the main topic. Subtopics can be chosen to narrow the search for assets. The Asset Management Switchboard created by the SW EFC also contains AM resources. The link for this switchboard is https://swefcamswitchboard.unm.edu. The switchboard can also be accessed from the main page of the SW EFC website (https://swefc.unm.edu/home/), selecting infohubs from the menu bar, and selecting Asset Management from the drop-down list. The switchboard is a collection of free resources available to anyone to help with asset management. The resources come from the Environmental Protection Agency, states, the SW EFC, the Environmental Finance Center Network, and non-profit organizations (Figure 8.4).

The information is sorted into five categories: What Is Asset Management, How Do I Get Started, The 5 Core Components, Benefits of Asset Management, and Beyond the Basics. The switchboard is updated periodically as new resources and tools are created or become available.

There are several other places one can look for resources. For example, the American Water Works Association has a resource page: AWWA: https://www.awwa.org/Resources-Tools/Resource-Topics/Asset-Management

Both AWWA and Water Environment Federation (WEF) offer conferences and events related to asset management. There are two main conferences that have a fair amount of content on asset management that are run jointly by AWWA and WEF, the Utility Management Conference and the Water Utility Infrastructure Conference. There are other conferences sponsored by these organizations that include asset management content as well.

In addition to AWWA and WEF, there are member organizations in the United States as well as international organizations that provide many

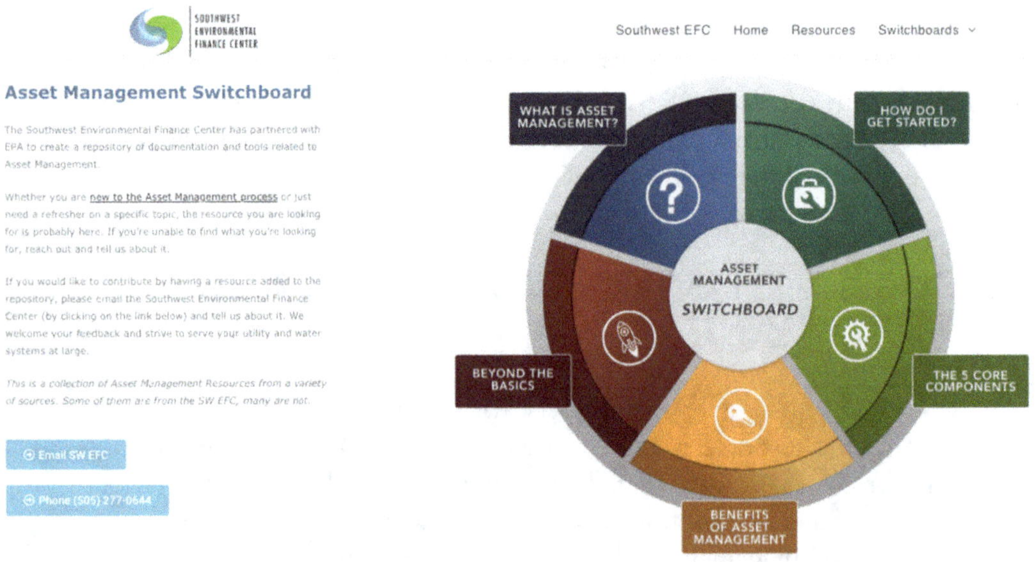

FIGURE 8.4 Asset Management Switchboard

resources, some free and some for a fee. Some of these organizations include the following:

ReliabilityWeb https://reliabilityweb.com/

Institute of Asset Management https://theiam.org/

Institute of Public Works Engineering Australasia https://www.ipwea.org/home

Inframanage https://inframanage.com/

Government Accountability Office (GAO) https://www.gao.gov/products/gao-04-461

There are also asset management resources available from many states, especially states that either mandate asset management as part of funding or incentivize it. Examples include Ohio, New Jersey, Oklahoma, Virginia, Michigan, and New Hampshire. There are definitely others.

The ReliabilityWeb holds a conference related to reliability and asset management that covers all types of industries, including water. It can be helpful to interact with others outside the water sector who are engaged in asset management to expand your thinking and hear different perspectives. Interacting in one way or another through conferences or just reaching out to others through phone calls or video calls is a good way to support your asset management learning journey.

11.0 REFERENCES

American Water Works Association. (n.d.). *Asset management.* https://www.awwa.org/resource/asset-management/

Southwest Environmental Finance Center. (2011). *Asset management kan work!* https://swefc.unm.edu/home/resource/am-kan-work-an-asset-management-and-energy-efficiency-manual-2/

Chapter 9

Maintaining Support for Asset Management

1.0 ASSESSING TRIPLE BOTTOM-LINE BENEFITS OF ASSET MANAGEMENT

The potential benefits of asset management fall into three categories—financial, environmental, and social—which are referred to as the triple bottom line. The possible benefits in each category are described below; however, the information is not an all-exhaustive list but rather presents examples of the benefits in various categories that may be achieved by developing and implementing an asset management program. While not every utility will receive all of these benefits, a well-crafted program is expected to achieve at least some benefits and over time, the benefits are likely to increase.

1.1 Financial Benefits

Financial benefits relate to monetary benefits and can result from efficiency gains or changes in practices. The diagram below shows the categories of financial benefits—cost savings, cost avoidance, and revenue benefits (Figure 9.1).

Cost savings occur when an activity continues to be done, but in a less expensive way, resulting in cost reductions. Some examples of this type of financial benefit are listed below.

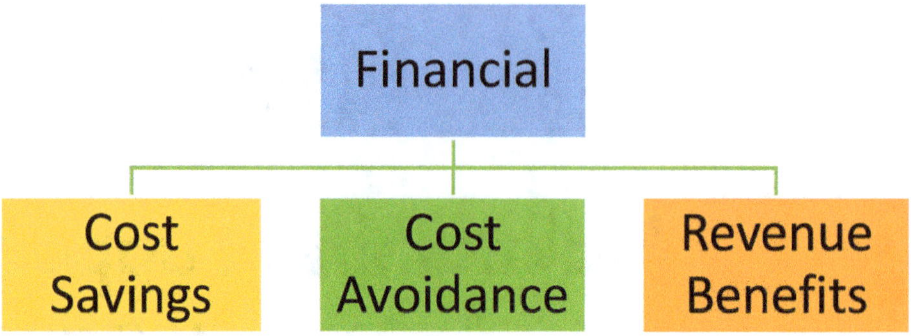

FIGURE 9.1 Financial Benefit Categories

- Reduced utility bills as a result of improved energy or water efficiency or by working with the utility company to address peak and off-peak usage.
- Work efficiency gains that reduce employee overtime expenses.
- Completing a rehabilitation project rather than a replacement project that decreases the overall construction cost.
- Reduction in expenses related to spare parts through the development and maintenance of a robust spare parts inventory system and improvements in dispensing and using parts in the inventory system.
- Reduced capital expenditures by extending the life of the assets through improved maintenance practices.
- Increased bond ratings from rating agencies which can lower the cost of borrowing for capital projects.

Cost avoidance results when changes in the operation, maintenance, or management of assets prevents something from occurring that would have otherwise. This cost is an avoidance, because it never occurred. It is similar to cost savings. Some examples of this type of financial benefit are listed below.

- Water efficiency practices reduce drinking water usage sufficiently to eliminate the need for a capital project (e.g., a new well).
- Strategic pipe replacement reduces the number of breaks in the system over time so fewer breaks occur and the cost to repair is avoided (select pipes or pieces of pipes with high likelihood of failure are replaced).
- Improved maintenance practices reduce asset failures and eliminate the cost of multiple repairs.

- Changes to individual maintenance practices extend the time between maintenance activities eliminating the cost of some actions (e.g., instead of replacing lubricant once per year, analysis shows it is only needed once every 2 years. The cost of one of the replacements is avoided).
- Changes to operation and maintenance of the facility brings the facility into regulatory compliance preventing a regulatory financial penalty or fine.
- An asset with frequent failures is improved rather than fixed (e.g., a plastic part that fails once per year is replaced with a stainless steel one that does not break for more than 15 years).

Revenue benefits occur when actions taken bring additional revenue into the utility.

- More customers are tied into the utility, thus increasing the number of customers who pay utility bills (e.g., connecting houses that were previously on wells or septic tanks).
- Utility increases the level of customer satisfaction such that customers are willing to pay higher utility bills.
- Inaccurate customer meters replaced so that bills more accurately reflect usage (can increase both the water and sewer revenues if sewer is based off water usage).

1.2 Environmental Benefits

Environmental benefits affect the surrounding environs—air, land, water, climate—and may require more effort to quantify than financial benefits. Initially, these benefits may be qualitatively assessed until the utility determines the best method to quantify them. The diagram below shows categories of environmental benefits (Figure 9.2).

Regulatory benefits are those that enable the utility to meet its requirements to protect human health and the environment. Some examples of these types of benefits are listed below.

- Increasing regulatory compliance, which can include going beyond what is strictly required by regulations (e.g., a wastewater treatment facility that reduces its biochemical oxygen demand and suspended solids to 10 mg/L each, even though the requirement is 30 mg/L).
- Reduction in the number of people receiving inadequately treated water or reduction in surface waters receiving inadequately treated wastewater.
- Reduction in the number of sewer overflows.

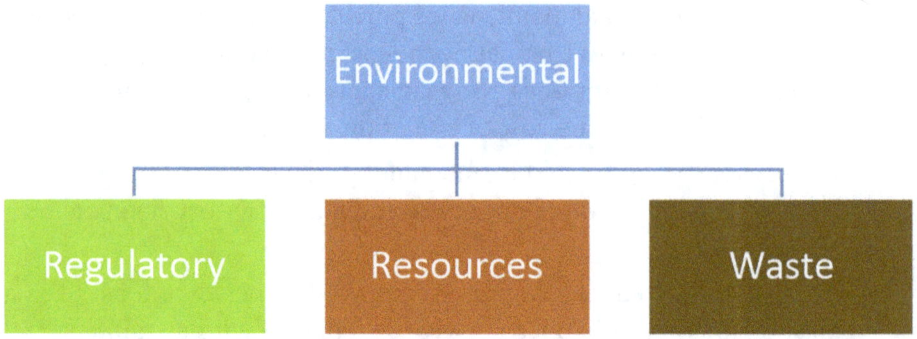

FIGURE 9.2 Environmental Benefit Categories

Benefits related to resources include those actions that would change the way resources are used or the type of resources used. For example, if a water-cooled mechanical device was replaced with an air-cooled device, it would reduce the use of the water resource. In a very arid region, this change can be quite beneficial. Other types of resource benefits include:

- Increase in energy efficiency (e.g., use of a more energy-efficient pump or blower).
- Potential to harness energy (e.g., use of an in-line turbine at raw water intakes to produce power for utility use or for sale back to the prime electrical service provider).
- Reduction in the system's carbon footprint.
- Change in energy usage from a more carbon containing source to a less carbon containing source (e.g., using biodiesel fuel or solar panels).
- Decreased use of resources in the treatment process.
- Allowing employees to clock in and out remotely at the beginning and end of shifts and/or lunchtime to eliminate the need to drive long distances to an office/work-site to clock in and out. (In some utilities with large service areas, considerable fuel may be used in transporting workers to and from time keeping equipment.)

Benefits related to waste include actions that will produce less waste or reduce the hazards of the waste produced. Examples are presented below.

- Reduction in quantity of waste produced in the treatment process (e.g., a more efficient reverse osmosis system that rejects less water).

- Reduction in the hazardous properties of the waste produced in the treatment process.
- Reduction in the discharge of waste products to the waters of the United States.

1.3 Social Benefits

Social benefits are those that affect people, including customers, elected leaders, and employees. Similar to environmental benefits, these benefits will take additional effort to quantify. As a starting place, they can be assessed qualitatively until a process to quantify them is developed. The diagram below shows categories of social benefits (Figure 9.3).

Asset management is, in large part, meant specifically to provide benefits to customers. One of the five core components—level of service—is aimed at providing the service desired by customers. Examples of specific benefits related to customers include the following:

- Increased customer satisfaction.
- Fewer service disruptions.
- Reductions in the number of customer complaints or in complaints of a particular type.
- Fewer blocked roads or fewer road cuts.
- More willingness to support rate increases or greater understanding of the need for rates.
- Fewer disruptions to businesses.
- Customers receive water or wastewater treatment who did not have such services before (or customers receiving adequate services who did not have adequate services before).

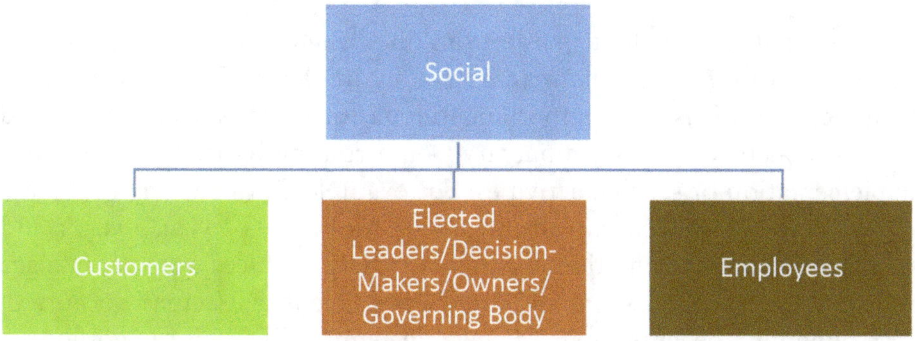

FIGURE 9.3 Social Benefit Categories

In addition to customers, some of the social benefits are meant to accrue to elected leaders/decision-makers/owners/governing body personnel. Examples of these types of benefits include:

- Better understanding of the need for overall funding for the utility or the need for individual projects.
- More public support for decisions.
- Reduction in number of complaints at public meetings.
- More supportive customers of rates, including accepting the need for rate increases.
- Public recognition of utility excellence and operational efficiency, incentivizing new businesses to consider locating within the utility's service area.

While the focus of social benefits is often on customers and elected leaders, it is important to remember that benefits should also be seen by the employees and staff. Since staff are instrumental in gathering data for asset management and carrying out the implementation of the asset management program, it is very important that they understand the ways in which the program benefits them. Some examples include:

- More planned, less reactive work.
- Fewer safety incidents and injuries.
- Increased job satisfaction.
- More ability to affect work and projects.
- Opportunities for career advancement.
- Potential for employee benefits such as pay increases or other monetary or non-monetary benefits.

The examples of the triple bottom-line benefits are just a few of the many potential benefits that can accrue by implementing asset management activities. It is most important to look for these benefits—large and small—monetary and non-monetary—on a routine basis, especially after implementation of specific activities. For example, if a new way of replacing water lines has been implemented, look for benefits 1 year, 5 years, and 10 years after implementation. If employee safety practices were put in place to help reduce accidents and injuries, check to see if that actually occurred. At a minimum, benefits should be examined annually in all categories.

The benefits achieved should be communicated widely to the public, the elected leaders, and the employees. Methods of communicating asset management activities and results are discussed in the next section.

2.0 COMMUNICATING ASSET MANAGEMENT

It is important to find ways to communicate the asset management program, level of service goals and ability to meet them, benefits (especially social benefits to customers), accomplishments and future expectations both within and outside the organization. There are many options for how to effectively communicate and each system/community should select the methods that work best for them. Likely, mixing several methods works best to reach as many people as possible. Some examples are discussed below.

- *Customer Conversations or Focus Group Meetings:* Customers can be asked to attend focus groups to share their opinions or provide specific types of feedback. During these sessions, the utility can share information about the asset management program, including how it is benefiting customers. The participants can be recruited through a variety of methods, such as mailers, bill inserts, social media announcements, e-mails invitations, announcements on the community web page, posters around town or in public places, such as libraries, or any other method the community has had success with or thinks will work. The attendance can also be incentivized, such as offering a discount on a future water or wastewater bill of a set amount (say $20 off). As another example, perhaps the next month the attendee is only charged the volumetric portion of the bill, not the base amount. When holding meetings of this type, the event should be highly interactive and engage the participants. It should not be a presentation, but rather a conversation that allows the utility to share information but also receive feedback. The utility decision-makers, including any elected leaders, should also be asked to attend so they can hear the feedback firsthand. If there are particular issues the utility is seeking opinions about, these should be the main focus of the meeting.
- *Using Media to the System's Advantage:* While the media is often viewed as the enemy or as only a source of negative attention, it is also possible to use the media in a positive way. A utility can reach out to the local media—either print, radio or TV—and ask them to share a positive story about the system and its asset management program. If the media is printing stories about some aspect of the utility, such as broken pipes, sewer overflows, or rate increases, the utility can take the opportunity to reach out to the media and provide asset management information, such as: "we are working on a capital improvement plan that will result in pipe replacement on a prioritized basis so we should see reduced break rates in the future. It will take us X years to complete the replacements." Or, "the rate increase is necessary

because we have under-invested in our facility for a long time and the increased number of sewer backups is a result. The rate increase will enable us to increase our sewer repairs, replacements, and cleanings." All media engagements should be handled by whoever is designated within the utility to speak on its behalf. In a large utility that might be the public information officer. In a medium utility it might be a public works director and in a small utility it might be a board member, or the head operator. Only one person should speak for the utility to make sure there is no mixed messaging. If there is a need for more than one person to engage with the media, all individuals should agree on the message that is to be conveyed.

- *Consider a Water Utility Resume:* A resume is meant to convey all of the accomplishments and positive aspects of an individual succinctly in a short amount of space. In a similar manner, a water utility can create a resume for itself to share all of its accomplishments over the past year or 5 years or 10 years (whatever amount of time is appropriate). Similar to the resume containing basic information about where someone went to school or what degree they earned, a water utility can describe how many miles of pipe they have or number of wells or number of customers. The resume can also include items such as: awards or recognition that was received, financial health, where funds were spent (major categories), construction projects completed, and similar. Figure 9.4 shows an example water resume.
- *Social Media:* There are many different types of platforms for social media. Some communities may use some social media sites more than others or different platforms may be used to reach different ages and types of customers. Information about the asset management program can be posted on social media during the year to inform the customers of the activities and successes of the program. If the community has an official social media presence on various platforms, the water utility may wish to have the asset management announcements added to these posts.
- *Videos or Podcasts:* With today's technology, it is easy to make videos and podcasts. A simple video or podcast can be made with a smart phone, tablet, or computer. These products can be made without spending much (or any) money and as much or as little time as the utility wishes to put into it. A few examples of topics for either video or podcast are: a day in the life of an operator, what it takes to make the drinking water (or wastewater) clean, why wastewater costs more than water, what testing is done to make sure the water meets standards or permit requirements, or where the money is spent. The videos

Water Utility Resume
(Your Water Utility's Story)

Basic Description:	Sources of Water Number of Miles of Pipe Number of Tanks Number of Valves, Hydrants, Meters
Major Accomplishments:	Replaced 20 Miles of Pipe Switched to automated meter reading devices Conducted leak detection on 10 miles of pipe
Financial Health:	Money in reserve accounts Financial Ratios
Funding Received for Capital Replacements:	SRF Loan for $2 Million
Awards:	Best Tasting Water Operator of the Year Award
Average Cost of Water:	Rate for an average water user in the community Comparison of cost of water to other types of products, including bottled water on a cost per gallon basis
Ability to Meet Goals:	Describe goals and whether each was met for the year
Your Water Utility Staff:	Describe utility staff (how many, qualifications, certifications) The daily duties/activities of operators

FIGURE 9.4 Utility Resume Example

or podcasts can be posted on the utility's website or social media or shared any other way that will reach the customers.

- *GIS Storymap:* GIS storyboards allow a utility to share its story, including its asset management journey. The pages can include the basics of the utility, level of service goal reporting, upcoming projects, current projects, completed projects, successes and anything else that can help customers relate to the work of the utility and the service it provides. The tools are relatively simple to use if a utility wants to create a storymap and it can share as much or as little as the utility wishes. It can also be expanded over time and grow as the asset management program matures.

In 2010, the Louisville Water Company was celebrating 150 years of operation and opened up its filtration plant to tours. They were not expecting very many customers to attend but were surprised to find thousands of customers wanted to see what the plant looked like. Because of the high interest, they held weekly tours for a year.

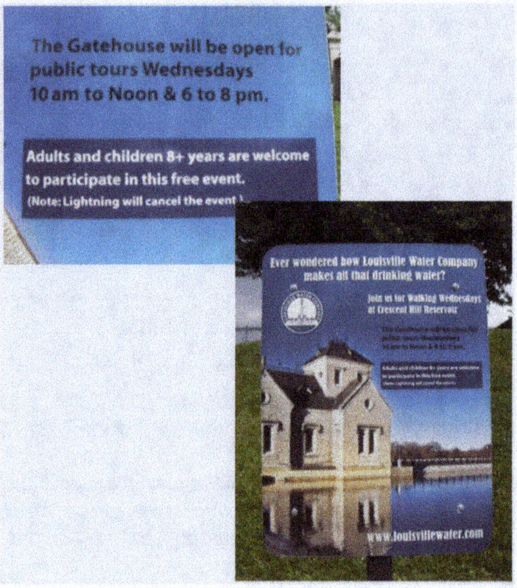

- *Plant Tours*: If the system has a water or wastewater treatment plant that can be toured safely or other affiliated assets, such as pump stations, storage tanks, or green infrastructure, a public tour can be a good way to encourage customer interest and enthusiasm about the work of the water utility. Many customers have no idea what goes into drinking water or wastewater treatment and can be surprised by how many assets are involved and how complex the process can be. If customers see what is involved in operating a utility, it may help them support the utility's efforts.
- *Bill Inserts*: If the utility mails bills to customers, bill inserts or statements on bills can be used to share information with customers on asset management topics. Many utilities are switching to electronic billing or automatic pay. In these cases, either the information would need to be mailed separately or it would have to be shared in one of the other ways mentioned in this section.
- *Consumer Confidence Reports*: Drinking water systems have to prepare a consumer confidence report each year and distribute it to customers. The report must contain some specific language, but it may also contain additional information that the system wants to share,

such as information describing the asset management program. It can include anything that the system wants to share about its asset management program, such as performance towards level of service goals, replacement projects, level of incremental funding, the long-term funding sustainability plan, or anything else.

- *Using Employees as Community Liaisons or Emissaries:* Employees of the system are frequently out and about in the community and see customers on a frequent basis. The employees can be given some training on ways to talk to customers so they can have casual conversations with them when they meet them during the course of their daily jobs. This type of interaction doesn't need to be on particular topics, just informal discussions. For example, if an employee is out removing a clog from a sewer pipe, they might engage in a discussion with the homeowner to say what types of things going into the sewer lead to clogs. They can indicate that "flushable wipes" are not really flushable and fats, oils, and grease lead to problems in the sewer. These problems lead to issues in the sewer system and can increase the cost of operating the system and shortening the lives of the assets.
- *Community Messages or Billboards:* For very large utilities, there are opportunities for billboards to convey messages. One great example is the billboards Denver Water used to convey messages around conservation. Other utilities, such as Albuquerque Bernalillo County Water Utility Authority have used billboards to convey asset management messages and messaged around fats, oils, and grease. While billboards might be out of reach for smaller sized systems, and some states do not allow billboards, there may be opportunities for systems to find other ways of sharing community messages. Possibly banners can be put up in downtown areas or posters can be created and put in places community members would see them. Electronic road signs can also be used to convey messages of importance, such as a major water main break or a sewer collapse. The town of Hanover, New Hampshire, used electronic billboards on the roads into town to encourage customers to complete their portion of the service line inventory required by the Lead and Copper Rule Revisions (LCRR) of 2021.
- *Engaging With Schools:* Another opportunity to share messages is through schools. Utilities can work with teachers at various levels to have them do lessons on water or wastewater, which can also potentially include guest speakers from the utility. These lessons can particularly focus on asset management related messages, such as what it takes to repair a system, the life of an asset, level of service and how that affects what utilities do, and risk assessment. When messages

are shared through school children, they often get conveyed back to the parents. The system can also sponsor contests around themes related to asset management that could include poster contests, models, Lego™ contests (Figure 9.5), and other related events.

There are of course many other ways to communicate with and engage customers and the methods are only limited by the utility's imagination. Several methods can be used at once to engage different types of audiences. The most important thing is to find some way(s) to engage customers. Customers are vital to the proper operation, maintenance, and finance of the system. This communication is necessary to sustain support for asset management within the customer base and ensure they understand what it is and why they should support it.

3.0 IMBEDDING ASSET MANAGEMENT WITHIN THE ORGANIZATION

Asset management often starts with or is driven by a "champion" within the organization who sees the value and wants to incorporate improved practices throughout the organization. However, it cannot stay with this one

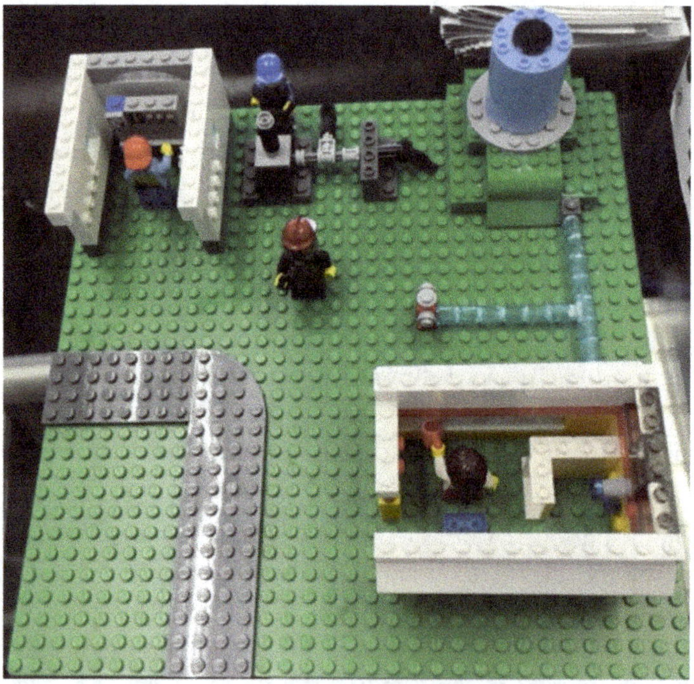

FIGURE 9.5 Child's Water System Lego™ Creation

person or these few people. It is vitally important that asset management become imbedded throughout the organization at every level, no matter how many levels that is. Asset management cannot be thought of as the "asset manager's job." It must be pushed down throughout the entire organization so that everyone knows what the program is and how it works.

It is necessary to ensure that the vision for asset management, as well as how it is being implemented, is communicated throughout the organization. One method of doing this is to ensure that all employees know what the mission statement is and can recite all or most of it from memory and understand how work performed (and measured via internal level of service goal targets) at the field/plant level directly relates to level of service goals conveyed to external stakeholders. Employees should understand how their actions enable a particular goal to be met and if the goal cannot be achieved, what actions can be taken to assist the situation. Employees should understand the progress towards meeting goals on a regular basis, whether that is monthly, quarterly, or annually.

All employees should have training in the basics of asset management and this training should be provided to any new employee when he/she/they start. If the system is too small to have an internal trainer, there are plenty of resources on-line that can be used (i.e., recorded webinars, podcasts, educational videos). If written materials are preferred, there are many of those resources as well that describe asset management at different levels of sophistication. As part of the training or after the training, it is important to ensure that each employee knows how the work they do directly affects asset management and the various ways in which their work intersects with the asset management program.

Having regular meetings with employees to describe the status of, challenges and issues with, and successes of asset management can be held. The frequency of the meetings depends on the size of the system, the number of staff, and the activities included in the asset management program but they should be no less frequent than annually. Large utilities may want to meet weekly or monthly. These communications should include as many staff as possible and if necessary, field staff can be rotated so that some field employees can attend each meeting but others remain in the field. One purpose of the meetings is to drive asset management practice as deep in the organization as possible.

Another way to stress the importance of asset management to all employees is to not only talk about asset management at the regularly scheduled meetings but to talk to employees at random times and ask them questions such as, "what does asset management mean to you?" or "What have you done to add to the program? What data have you collected and how did you collect it?" It is also important to ask if there are any concerns with how the program is going or if the employee has any questions that can be addressed. Employees should feel empowered to share any and all concerns about the process, the goals, progress, or how to handle the associated changes.

Asset management practice is improved by asking questions and accepting that just because you have always done something one way does not mean it always has to be that way. The questions are likely to come from front line staff and one sign that asset management is being imbedded into the organization is when these employees feel empowered to come to management and share their questions. Management needs to listen to the questions and take them seriously. The questions can be explored to see if there is a better way of doing something than the way it is being done or if some other change is worth pursuing. If employees see that they can ask questions, the questions are taken seriously, and some positive changes result, the successes will drive employees to support asset management and share their support with each other.

To sustain asset management practice, it must be possible for the "champion" to leave or retire without causing major problems in the implementation. The program should continue forward even when personnel are replaced. Ideally, a succession program exists to allow employees to be replaced seamlessly within the organization.

4.0 THE HUMAN ELEMENT

Because asset management is a thought process instead of a technological process, people matter. They are the most critical element in whether or not asset management succeeds. The staff need as much attention as the physical assets in the system, whether or not they are called "assets" themselves. They require training—both classroom and in the field; mentoring; support from leadership; opportunities to learn at state, local, or national conferences; the proper tools to do their job; tools and resources to ensure their health and safety; and a salary commensurate with their skill set and experience. They also need to understand the staffing plan through the entity's organizational chart, have support to fill vacancies when they arise, and the opportunity to seek additional employees when the workload demands additional help. The employees should also be able to have a career development plan to understand how they can grow in the organization and a succession plan when they approach retirement.

When a utility decides to embrace and implement an asset management program, there will be change. Managing this change is necessary to allow individuals to understand their role in the program, reap the benefits, and have input into how the program is developed and continued over time. Employees must be empowered to be part of the solution and to share their wisdom and knowledge. Without good people, asset management will fail or at most will only achieve a small part of the potential benefits. It is critical for upper management and elected leaders to recognize the value of their employees and invest in their success, because that is an investment in the success of asset management overall.

The thought process of asset management means that many of the decisions regarding what to do with a particular asset rely on the employees. For example, there might be enough money to replace one mile of pipe. Someone must decide which mile should be replaced. Using asset management principles means making the best decision possible to ensure the lowest life cycle cost. Employees must use all the available information to determine the best course of action.

This seemingly simple question (which mile of pipe to replace) requires numerous employees to do their jobs, including: collection of break data by field crews to provide information related to probability of failure of the pipe; field crews completing a condition assessment for pipes they repair; individuals creating an inventory of pipes including attribute data; individuals creating a map of pipes and break data so the data can be presented visually; individuals field verifying data to ensure its quality; individuals determining a means of assessing consequence of failure of pipes; individuals analyzing all of the data to determine the criticality of the pieces of pipe in the system; and a decision-maker making the ultimate choice of which pipe to replace. Having people involved in the process at every level is absolutely vital to implementing asset management. (In a small system, employees would likely be doing multiple activities from this list rather than having different people performing these functions.)

People can fall along a spectrum from those who oppose asset management and cannot be convinced that it is a worthwhile process all the way up to asset management champions. The graphic below shows the spectrum (Figure 9.6).

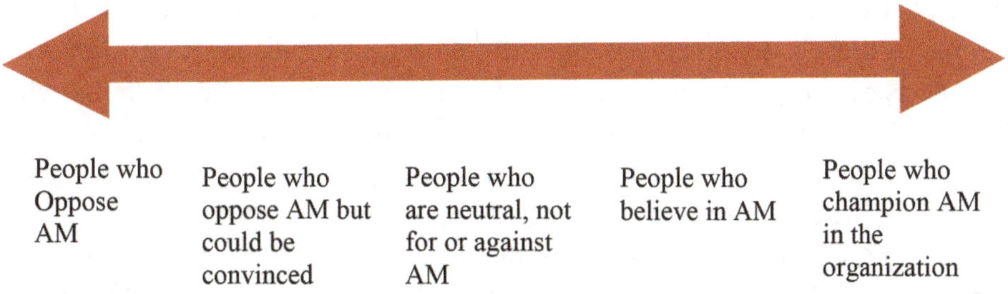

FIGURE 9.6 Asset Management Acceptance

The people on staff who oppose asset management and cannot be convinced otherwise can harm the overall efforts to implement a program. These individuals may need to be removed from the organization if they get too strong in their hostility or start to sabotage efforts. These individuals may not be honest about their feelings, which can make it even harder to succeed. These individuals may be thought of in a similar way to a failing asset. A failing asset needs to be removed and replaced in the system in order to prevent negative consequences of failure. Similarly, an employee that is not willing to learn or grow with the utility's implementation of asset management can cause negative consequences if not addressed.

The next set of people are those who are opposed to the concept of asset management but if shown enough success or benefits for either themselves or the organization can be brought around. Generally, these individuals are up front about their skepticism and just want to see evidence of the benefits before getting on board. These individuals can eventually be won over and may even end up being some of the biggest champions.

Individuals who are neutral may see some potential benefits in asset management but do not know if it is worth the effort. There is potential to eventually move them in the direction of supporting asset management efforts. It may require a bit of patience to allow these individuals to grow with the program and see the positive aspects.

People who believe in asset management from the beginning are valuable to the organization. They understand what implementing an asset management program can do for the utility and will openly support the efforts. They can also help encourage other employees to embrace the movement towards improved asset management practice.

The final group are the champions for asset management. These individuals strongly believe in the principles of asset management and will do what it takes to convince others to join in the program. It is harder to implement asset management without at least one champion. Otherwise, there is a lack of passion and enthusiasm which makes it more difficult to sustain the program through difficult periods.

The hope is to have more people in the last two categories than the first two. Fortunately, with all the talk of asset management, there are fewer people in these categories now than there used to be. The organization can also consider sending some staff to training or to visit other systems who have implemented asset management successfully to try to build staff support for asset management before starting the journey.

While the champions are extremely important to asset management, it is critical that everyone sees asset management as part of their job. Everyone plays a role and it is important that individuals see it that way. This component, as well as several other aspects of the human element, are shown on the table below along with a discussion of metrics that can be used to help evaluate them (Table 9.1).

It is also important for employees to understand the overall benefits in implementing an asset management program and as well as the specific

TABLE 9.1 Human Elements and Measurement Metrics

Human Elements	Metrics
Asset Management Is Everyone's Job	Is asset management included in everyone's job descriptions? If not all, which ones include it and which ones do not? Why is it not included? Can it be added to all job descriptions over time? If it is not specifically stated in the written job description, is it at least a verbal understanding? Is asset management one of the ways employees are evaluated during employee reviews? If yes, in what ways? If no, why not? Can this be part of reviews in the future? Do individual employees have asset management goals they need to meet monthly and annually? Would an individual employee be able to state how he/she/they fit into the asset management program?
Buy-in Must Be Throughout the Organization	Would every employee be able to define asset management if asked? Is the basic asset management terminology understood throughout the organization? If one person leaves the organization, will someone else be able to seamlessly fill in without affecting the asset management program? Would everyone be able to describe, in their own words, the "what" and "how" of the asset management Program?
Knowledge Management & Succession Planning are Important Components	How is employee knowledge captured on a daily basis? What is done to capture the cumulative knowledge of long-term employees? Are critical jobs identified? Do critical jobs have a succession plan? Is it possible to do cross-training to improve work efficiency and offer career advancement?
Employees Should Be Empowered to Suggest Changes Based on Their Experiences and Backed Up By Data	Number of activities suggested by employees How many activities suggested by employees were initiated? Benefits associated with employee suggested activities (benefits should be shared widely to inspire more work of this type).

benefits for them. The benefits are not just for others—customers, elected leaders—but are for the individual employees as well. While there are many potential benefits for employees, a few examples are shown below.

Safety improvements	More planned activities during the day; less reactive	Ability to "fix" problems not just band-aid the situation	Opportunities for career advancement
Empowering employees to suggest and/ or design asset management activities	Potential for employee benefits such as pay increases or other monetary or non-monetary benefits	Less "busy work"—greater understanding of how their work matters	Increased job satisfaction

It is not necessary to wait to consider potential benefits; employees should be encouraged to think about benefits they would like to achieve at the very beginning of the process so that these activities can be built into the strategy.

It is also important that the employees understand that asset management will require some change. It is not possible to achieve different results while doing everything the same way. However, to reduce the concerns somewhat and ease the transition, asset management can be designed to address some of the biggest needs of the organization first to get buy-in during the transition. It can also be designed to work as much with the existing culture of the organization as possible. For example, if the organization is one that does not rely heavily on technology, simpler approaches can be used or approaches that reduce the number of people who have to use the technology can be undertaken, especially in the beginning. For example, maybe all data is initially entered on paper and then one person, rather than everyone, enters the data into a computer program. The main objective is to build asset management into the system in a way that most employees feel comfortable with it and are willing to embrace.

To help ensure that employees are supported throughout the development and implementation of a robust asset management program, the following process may be helpful: (1) Develop a staff champion, (2) find a utility leader who can supply the necessary resources, (3) develop an improvement program communication plan, (4) start small, (5) scale up over time, (6) create a multi-year vision, (7) determine a continuous improvement path, (8) develop a comprehensive approach by adding elements, and (9) sell your approach! While this discussion of staff is found near the end of the book, it is important to consider staffing as one of the key elements of a successful program and to ensure that staffing components are built in throughout the asset management program development and implementation.

Chapter 10

What an Elected Leader Needs to Know

Elected leaders, governing bodies, volunteer water leaders, or owners are entrusted with making decisions for their water systems including approving and funding necessary replacement projects, hiring executive level employees, setting the overall mission and direction, and monitoring progress towards meeting goals. These leaders may be responsible for all community assets or just the water assets. In either case, it is possible that water is not the primary field of knowledge for the leaders. Water is complicated and may represent the single biggest investment for a community. It is also largely out of sight, buried infrastructure so it is much less front of mind than roads, parks, libraries, pools, emergency services, or other functions.

Drinking water pipelines can cost $1 million to $3 million or more per mile to install and wastewater can be double or triple that cost. A drinking water system with 1000 miles of pipe could easily cost over a billion dollars to replace. As an example, the assets in the Albuquerque Bernalillo County Water Utility Authority were valued at approximately $5 billion dollars, including all the pipe and treatment assets for the water and wastewater systems during their 2011 Asset Management Plan creation. The assets would even be more expensive today (Albuquerque Bernalillo County Water Utility Authority, 2011).

In this section, we will refer to these ultimate decision-makers as "elected leaders" but it should be understood that this terminology refers to any type of water leader whether a volunteer, owner, or other decision-maker. An entire book could be written on leadership for water utilities, the use of data-informed decision-making for leaders, and leadership related to asset management. Such an extensive foray into leadership is outside the scope of this book. However, there are resources from organizations, such as the Institute for Asset Management, the Federation of Canadian Municipalities funded by the government of Canada, the Government Finance Officers Association, and many others that can provide additional information for elected leaders to learn more.

One of the best ways for an elected leader to get started on their journey towards understanding the water systems (drinking water, wastewater, and/or stormwater) is to lay eyes on the assets. For small communities, and possibly even larger ones, the investment in water infrastructure may constitute one of or even the biggest investment of capital. There may be much less awareness of this infrastructure given that much of it is out of sight. It is hard to imagine how leaders can make good decisions about asset funding, rate increases, operations, management, budgeting, or anything else without a good sense of the breadth, depth, and complexity of the facilities. It is highly recommended that elected leaders take the opportunity to visit the visible assets (e.g., treatment plants, well houses, pump stations, lift stations, storage tanks, stormwater conveyance, outfalls, manholes, and raw water sources) and get a sense of the condition, the useful life remaining, the needs for maintenance, and any other pertinent information.

The leaders should also get a sense of the amount of piping in the ground by driving around the area and/or looking at maps of the distribution, collection, or stormwater conveyance systems. This can be tens of miles in a small community, hundreds of miles for a medium community, and thousands of miles for a large community. All of this infrastructure must be managed, repaired, and replaced over time.

There should be at least one visit to facilities by every leader but preferably, the visits would take place annually. Annual visits are a chance to observe how the assets have changed over time. As new leaders come on board, they should have a tour of the facilities as soon as possible. This type of firsthand knowledge can be illuminating and help convey the sheer magnitude of the investment and the need to carefully consider rates and revenue requirements. It is hard to make good decisions about replacement or rehabilitation if the assets have never been seen and there is no sense of what the overall condition is.

In addition to visiting the facilities, it can be helpful for elected leaders to spend some time with water, wastewater, and stormwater operators and managers to gain a sense of what an average day entails. Even spending an hour or two riding around with an operator can shed light on what the major tasks are and how important water is to the overall community.

Because implementing asset management practice entails meeting the desired level of service at the lowest life cycle cost, it is one of the best tools for leaders to implement in their communities. Ideally, they would use asset management for all types of assets within their purview, but at a minimum, it should be practiced for water infrastructure. To ensure that the asset management program has the best chance of success, elected leaders must demonstrate support for it at the highest levels. This can be done by passing a resolution, or its equivalent, and/or creating an asset management policy statement to indicate that asset management is the way business will be done. The support should also include a level of investment in asset management on an initial and ongoing basis to support personnel dedicated to asset management and any necessary technology purchases (software and hardware). For asset management to achieve its full potential benefits, to continue sustainably, and evolve, it must be imbedded in the organization from top to bottom.

The resolution should include some or all of the following:

- A description of the assets that the resolution applies to
- An acknowledgement that the system will be practicing asset management (or is practicing asset management)
- Any other related policies (e.g., sustainability) that should be acknowledged
- An approval statement for the resolution
- Any other information the entity wishes to include

There are many resolution examples online for asset management that can be used as starting places to craft an asset management resolution.

In addition to or instead of the resolution, an asset management policy statement can be created. The policy statement can include some or all of the following:

- Purpose statement
- Policy statement
- Policy objectives
- Policy principles or guiding principles
- Introduction
- Scope
- Application of the policy
- Overview
- Potential outcomes
- Associated regulations and standards
- Definitions, acronyms, and abbreviations
- Asset management team

- Asset management leadership
- Attributes

The asset management policy statement should be no more than a few pages long and should use easy-to-understand language so that anyone can read and comprehend it. Similar to the case with resolutions, there are many examples of asset management policy statements online that can be used as starting points and modified as desired to meet the individual needs of the water utility.

While it is not necessary for elected leaders to be involved in all the details of asset management, there are some portions that are key for them to understand and potentially be involved in. Descriptions of these items are presented below.

- *Mission Statement:* The organization's mission statement should be created and/or approved by the elected leaders or at least sanctioned by them. They need to understand that the selected mission statement will identify which actions the organization needs to prioritize within the budget.
- *Goals:* Elected leaders need to approve and/or sanction at least the external goals. Internal goals do not necessarily need to go through approvals as these goals are related to meeting the external goals.
- *Goal Targets:* The goal target levels (at least the external goal targets) should be acceptable to the elected leaders and the external levels should be approved at a regular meeting.
- *Progress Towards Meeting Goals:* There should be regular reporting of progress towards meeting goals in a manner that is easy to understand. The frequency of presenting the progress depends on the particular goals. Most likely the reporting would be at regularly scheduled meetings with some goals presented at every meeting while others are presented quarterly, semi-annually, or annually.
- *Action When Goals Not Met:* One of the main purposes of having and measuring goals is to allow for the discussion regarding what happens when goals are not met. Sometimes the actions might be simple, while other times, budget revisions, rate increases, staff augmentation, equipment purchases, or asset replacement may be required. When the action requires a leadership decision, the elected leaders need to be prepared to act.
- *Assets:* While it is not necessary for elected leaders to understand every asset in the facility along with their associated attributes, it is important to gain a sense of the magnitude of the assets—how many

in total, what types, their overall condition, and the expected useful lives and how that relates to the need for replacement. This type of knowledge can be gained from some well-crafted tables, graphs, and charts that show summaries of information, rather than long lists of hundreds, thousands, or tens of thousands of assets. If the system has an electronic computerized maintenance management system or asset management system software, elected leaders may be able to have a dashboard suitable to their needs for general information about the system. The dashboard might include total numbers of work orders and how many are open or closed at that given moment as well as a ratio of planned work orders to corrective work orders. Depending on the sophistication of the program, it can also include information on risk, useful life remaining over time, condition across the organization, operational data, or any other information the individual finds useful in general decision-making and oversight of the facility.

In the absence of a dashboard, it is also possible for the manager or operator of the facility to develop a simple "state of the assets" report or presentation or handout for the elected leaders. The state of the assets can include items such as: information on risk, useful life remaining, condition across the organization, operational data, and some performance measures with metrics. This information can help spark discussion between staff and leadership around assets and the need for resources to address specific issues, concerns, or assets.

- *Asset Map:* Elected leaders should be able to access a map of the facility in either paper or electronic format. The map can be printed out and posted on the wall to show overall size of the service area and other pertinent information and/or it can be a geographic information system map that can be easily accessed.
- *Understanding Criticality:* One of the best ways to have more effect with the same amount of money or to have the same effect with less money is to understand the power of criticality. It is very important for elected leaders to ask questions regarding whether the money is going to those assets that have the highest criticality. If the money is not going there, it is probably not going to the correct asset(s). It is relatively easy to craft some form of criticality visual (as discussed in Chapter 4) that can be shared with leaders to present the overall criticality of the organization.
- *Risk Tolerance:* In any set of assets, there is a given risk tolerance based on the condition, the probability of failure and the consequence of failure, whether it has been quantified or not. Asset management attempts to help quantify this business risk to allow the ultimate

decision-makers to address it. Risk and cost are interrelated in many complex ways so it is a balance to determine the risk tolerance. Too much risk means that bad events may occur, some of which may be unacceptable to the community. A very low risk profile may require a level of investment that is beyond the financial capabilities of the community. Therefore, risks must be tolerated, but at a level that is acceptable to the community. It is important for elected leaders to understand what the current risk profile is and to make decisions about whether the risk fits the risk tolerance of the community. If not, additional investment in asset interventions (operation, maintenance, repair, rehabilitation, or replacement) is likely required.

An option for elected leaders to consider is the development of a risk management policy that can specify how risk is to be factored into decision-making. The policy can specify actions that must be taken based on a given level of risk. The risk policy may state who the risk should be reported to, whether the risk is tolerable or not, what actions should be taken and how soon they should be taken. A simple example table that could be included in a risk management policy is shown in Table 10.1. The table can be made more detailed to meet the utility's needs.

- *Understanding Asset Life Cycle:* It is a good idea for elected leaders to have a basic understanding of the overall asset life cycle. It is important for leaders to realize that an asset's life starts during planning and design and the importance of including operation and maintenance personnel in this phase. Elected leaders can help ensure that this cooperation and partnership occurs. Elected leaders can ask questions of the managers and engineers (both internal and consultants) to judge the level of collaboration and if it is not sufficient, to work towards increasing it.
- *Corrective vs. Planned/Proactive Maintenance:* While it would not be expected that an elected leader would understand a lot about maintenance, especially specific maintenance tasks, it is expected that a leader could understand the ratio of planned maintenance to corrective maintenance. The further the ratio skews to corrective or reactive maintenance (only addressing assets after they have failed), the more likely the facility is spending too much money on maintenance. The elected leaders should ask the managers to provide a status of reactive to planned maintenance at least annually to ensure that the ratio is trending in the right direction.
- *Capital Improvement Planning:* Elected leaders should expect the utility personnel to develop short- and long-term capital improvement

TABLE 10.1 Example Risk Management Policy Action Table

Risk Rating	Report To	Tolerable Level	Actions	Timeframe
Low POF/ Low COF	N/A	Tolerable	Action is unlikely for these assets at this time.	N/A
Moderate w/ High POF Low COF	Supervisor	Tolerable	Evaluate whether there are cost effective strategies to reduce risk or replace assets. Potentially perform condition assessment activities if cost effective.	2–10 years
Moderate w/ Low POF High COF	Asset Manager/ Elected Leaders	Conditionally Tolerable	Detailed condition assessment conducted. If repair/ rehab/replacement needed, develop business case to support project.	2–5 years
High POF/ High COF	Asset Manager/ Elected Leaders	Intolerable	Propose repair/ rehab/replacement options to reduce or eliminate risk.	Immediate to 1 year

POF = probability of failure; COF = consequence of failure; N/A = not applicable

plans. It is necessary for elected leaders to understand the big picture regarding asset replacements and when those replacements will be needed. Without that big picture it is easy to push off investments and imagine that it is okay not to invest for a year or two. If the long-term plan is understood, the investments cannot be put off because there is always something else that needs to be funded.

- *Incremental Investment:* Elected leaders need at least some understanding of the utility's need to make incremental investments, particularly in large dollar assets, such as pipe. Incremental investment will help ensure that the community's long-term needs are being considered and that the utility will be sustainable over the long term. Having poor water infrastructure that is greatly underinvested in is not going to sustain the community.
- *Rates:* The rates are the ultimate responsibility of the elected leaders. Others within the utility can suggest the amount of revenue needed

and even a particular rate structure. However, the decision to enact a particular rate, raise a rate based on need, or raise a rate based on a set factor or percentage rests entirely with the elected leaders. It can be hard to take that vote but if the customers understand the need for the funds, it can make it a little easier. One aspect of asset management is that it is intended to increase transparency—to enable customers to understand what service level is being provided and how well that is being accomplished. With this increased transparency, along with communicating clearly regarding the need to replace assets, customers may have fewer objections to rate increases. In any case, for the good of the community, it is often necessary to find the courage to raise the rates to meet the needs of the facility, even if complaints will occur.

- *Outside Funding:* Any water or wastewater facility is likely to need at least some outside funding to address capital investments in infrastructure. Elected leaders need to approve this funding and may need to help with completion of applications or application requirements and possibly to approve expenditures for outside professionals. Elected leaders should request to see comparisons between funding sources to make sure that the selected fund(s) are most appropriate. It is also important that the correct alternative is chosen for the community. The elected leaders should ensure that a robust process has been undertaken, that staff have been involved in the discussions, and that the whole life cycle has been considered, not just capital costs.

Elected leaders are given the awesome responsibility of choosing how to spend the limited resources of the community. They cannot make the best decisions without truthful, transparent information. There may be a temptation to match the request to the budget rather than making the budget match the request. What this means is that the request coming forward has already been reduced to match what the personnel expect to receive, but this adjustment comes at a cost. There may be considerable risk imbedded in that budget. For example, consider a utility that has more than 80 wells and has not invested in wells for over 20 years. The profile of the wells is such that the condition of most wells is 3 to 5 (average to poor) with only a few rating a 2. The need for new wells is urgent and the expected number is five to six wells. However, the wells carry a cost of $2 million to $4 million each from initiation to completion, which will take approximately 3 years. Without the new wells, likely water shortages may occur and water quality may deteriorate, possibly below regulatory values. The budget for an investment of this type is $1 million per year currently, so only one well could be drilled in the 3-year period. One option is for the utility personnel to take a project to the elected leaders that includes only one well and does not address the risk associated with the need for four to five more. On the positive side, the investment is within budget; on the negative side, the risk is being hidden from the very people who need to understand it the most. This type of situation also works against the goal of transparency in asset management.

If instead, the personnel bring forth a request to have five to six wells drilled at a cost of $10 million to $24 million with all the backup data to show: (1) the need for these wells, (2) the risk if the wells are not constructed, and (3) the possible negative consequences to the community, the elected leaders can understand the risk and make an informed decision. They may still only agree to one well and choose not to invest more money than the current budget allows. However, now the risk of that decision has been taken into account and they made their decision with full knowledge of this information. It is important for utility personnel to share these hard truths about the facility and its needs for asset replacement and management. If the benefits of asset management are to be realized, it is necessary to be transparent with information.

Furthermore, if the elected leaders are aware of all infrastructure needs, they can choose to pursue outside infrastructure funding to cover the costs and complete the project that way. The amount and type of funding varies from year to year so if elected leaders choose not to seek funding immediately, because they are aware of the needs, they can be prepared to apply for funding if and when more favorable sources are available. In 2021, the Infrastructure Investment and Jobs Act (more commonly called the Bi-Partisan Infrastructure Law or BIL) was passed which allocated $50 billion over 5 years into the State Revolving Funds for drinking water and wastewater. A portion of this money was intended to be in the form of principle forgiveness (similar to a grant) for disadvantaged communities and many states also lowered their interest rates for loans of this money. A community that is prepared with the knowledge of all of their infrastructure needs is in a better position to apply for and, potentially, receive the money.

As leaders turn over, the new leaders need to be educated and informed of the asset management program and their individual role within it. It will be difficult to sustain the asset management practice without this transfer of knowledge to new leadership. The resolution, asset management policy document, and asset management plan (if one has been developed) can be used as part of the process to educate new leaders about asset management. However, field visits and in-person meetings to discuss asset management practices, successes, benefits, and challenges should be set up to educate the new leaders.

For asset management to be as successful as possible, the involvement of elected leaders is a critical component. Utility personnel should work to ensure these individuals are being kept informed about asset management in general and the specific needs of the system. While it is possible to perform some asset management tasks and begin a program without leadership support, there will eventually be roadblocks to increasing the practice. The roadblocks may be related to a lack of funding for a particular asset replacement, software, hardware, or other necessary purchase; lack of support for staff changes, increases, or particular positions; or difficulties related to getting cooperation from different parts of the organization. If it is necessary to start asset management related activities without the full buy-in of the leadership, it is necessary to look for ways to share successes and try to bring the leaders on board over time.

1.0 REFERENCE

Albuquerque Bernalillo County Water Utility Authority. (2011). *Asset management plan written by GHD*. https://www.abcwua.org/wp-content/uploads/Finances_PDF/2011_Asset_Management_Plan.pdf

Appendix A

AMP Outline

This outline offers one option for an asset management plan (AMP). There are many other options and possibilities, and this outline can be adapted to meet the needs of the entity using it. It is best to keep the AMP as succinct as possible without a lot of background on asset management in general or other general information. Also, it is best to keep very specific asset information out of the document to prevent it from becoming obsolete too quickly. The document should be able to remain relevant and current for approximately 5 to 10 years.

I. EXECUTIVE SUMMARY

Relatively short section highlighting the asset management issues you would want to convey to elected leaders and other interested parties who are unlikely to read the entire document (or potentially do not need to read the entire document). Should not have any new material that is not elsewhere in the document.

II. INTRODUCTION

This section should be an introduction to what the document is about, what the AMP development process was, and who was involved. If desired some of these items can be expanded and put in their own sections.

a) Commitment of the organization to asset management
b) Short description of asset management
c) Purpose/scope of AMP
d) Structure and content of the plan
   Describe the structure and contents as well as how often revisions are expected.

e) Vision, outcomes, and priorities for the AMP

f) Development process

Describe the development process for the AMP. What meetings were held, who was involved, over what period of time, whether an asset management committee was formed, who wrote the plan, how it was written, and any other pertinent information about the asset management development process.

g) Equity issues

Describe any equity issues, concerns or equity-related items for the AMP and its implementation.

h) Relationship to other plans, policies, documents, or strategies

Describe how the AMP relates to any other plans, policies, documents, or strategies that may exist within the organization, community, or larger organization the utility is located within.

i) Key stakeholders

j) Key challenges over the next decade

What are the key challenges the organization must address over the next decade (or other length of time) that relate to the development and implementation of the AMP and program.

k) Opportunities over the next decade

What are the key opportunities available to the system over the next decade (or other length of time) that relate to the development and implementation of the AMP and program (for example, IT, funding, software, other).

l) General description of overall facilities/assets

   a. Service area

   b. # of assets

   c. Value of system/replacement costs

   d. Age/history of the system or department or entity completing the plan

   e. Materials of assets, especially pipe assets

   f. System physical capacity

III. DEMAND—CURRENT AND FUTURE

Describe the current demands on the system as well as the future demands. The future demand considerations can be extremely important if they are expected to change a lot (either positively through growth or negatively through reduction in population) and affect the overall asset inventory and asset operation, maintenance, and repair/replacement. If the future demand is not expected to be tremendously different than current demand, this section can reflect that not much is expected to change in the system over the next decade or so.

a) Description of current demand requirements
b) Demand drivers

   Demand drivers can be items such as new development, serving unserved areas, climate change increases or decreases, asset performance (poor performance of assets could result in changes in demand), population growth or decline, tourism increases or commercial or institutional customer changes.
c) Expectations for future demand as well as how it was determined
d) Expected changes/revisions to meet future demand
e) Any assumptions associated with projecting future demands
f) Approaches that will be used to manage future demand (asset and non-asset based approaches)

   Non-asset approaches could be something like educating the public on what to and not to put down the toilet, or how to dispose of fats, oils, and grease (FOGs), distributing FOGs collection containers to customers, encouraging customers to install local green infrastructure to infiltrate stormwater, etc. Or on the water side, encouraging customers to conserve water to reduce demand so that a new well is no longer required.
g) Financial impacts

IV. LEVEL OF SERVICE

This section includes the mission statement as well as a description of how it was developed. It also includes level of service goals, how they will be measured, how often they will be measured, who will measure them, what data is currently available to measure goals, what data needs to be added and how it will be added, how goals will be communicated and to whom and how often.

a) Mission statement
b) Level of service goals, including target levels for each goal

If the target levels are expected to change over time, the current and future levels should be described as well as when the levels are expected to change.

c) Engagement with customers to understand their needs

d) Level of service goal measurement

If the measurement includes any type of electronic dashboard or visual of some type (electronic, paper), describe the dashboard or visual that should be used.

e) Data needed to measure goals as well as where to get it and what needs to be added if it does not currently exist

f) Level of service goal communication

Who will the goals be communicated to? How often? In what ways?

g) Actions to address goals not being met

V. CURRENT STATE OF THE ASSETS

This section should include information about the assets such as: what asset classes are included in the inventory, where the inventory is stored (e.g., Asset Management software, computerized maintenance management system [CMMS] software, and/or geographic information system [GIS]), what asset attributes or data is included in the inventory, and how the data will be used (i.e., what data categories are used and by whom).

a) Asset classes
b) Asset ID assignment
c) General description of number of assets in inventory
d) Asset hierarchy
e) Attribute or other data about assets and asset classes included
f) Mapping of assets/asset mapping information
g) How data will be gathered about new assets
h) Green assets defined as assets, if green assets are part of the utility/system
i) Asset data analysis (how data is used and by whom)
j) Missing data attributes and how they will be added to the inventory
k) Assessment of data confidence—Overall assessment of asset inventory completeness
l) Asset condition assessment—Description of condition assessment process for all asset classes
m) Summary of asset condition

n) Ways of assessing useful life remaining

o) Summary of useful life remaining

p) Asset inventory software(s) used

Include all types of software used, asset management (AM) software, GIS software, CMMS, Enterprise Asset Management System, or any other type of software.

q) Future changes planned or needed to asset inventory software

r) Asset ownership and its effect on operation, maintenance, repair, replacement

If asset ownership is mixed (assets owned by private entities, other municipal departments, other governmental entities, or a combination of all), discuss how those ownership types affect future operation, maintenance, repair, replacement, or any other part of managing assets, including funding.

VI. CRITICALITY

This section includes information about the criticality of the assets—the probability and consequence of failure for assets, the associated criticality, and how they map onto a criticality chart or other visual.

a) Failure modes

b) Methods of rating probability of failure (POF), either as a whole or by asset class if different for different classes

c) Methods of rating consequence of failure (COF), either as a whole or by asset class if different for different classes

d) Calculating criticality (POF x COF)

e) Processes used to reduce risk

f) How redundancy is included in analysis

g) How risk is visualized

h) Summary of risk visualization

i) Overall risk tolerance

j) Whether risk profile matches risk acceptance by decision-makers

VII. LIFE CYCLE COSTING

This section includes information about the full life cycle of the assets and the types of interventions that can occur at each stage.

a) Description of overall asset life cycle

b) Goals with asset life cycle management

Examples of goals that may apply to asset life cycle management include:

- Provide and maintain assets economically and deliver services affordably.
- Achieve the asset's service potential.
- Achieve customer levels of service.
- Achieve health and safety standards.
- Reduce organization's exposure to risk from unforeseen asset failure.

c) Description of each stage of the asset life cycle and the types of interventions that can occur

Include staffing implications for interventions for each component of the life cycle:

i. Planning
ii. Design
iii. Construction
iv. Operation & Maintenance
v. Repair
vi. Rehabilitation/Replacement
vii. Abandonment/Disposal

d) Using interventions to extend asset life or make other improvements

Include a determination of which interventions to do based on available resources.

e) Description of proactive vs. reactive or corrective maintenance

If there is a desired future condition that is different than the current condition, that should be discussed along with the methods/activities to reach the future condition.

f) Creation of operation and maintenance plan

If there currently is no operation and maintenance plan, will one be created in the future? If so, how, by whom, and by when? Will the O&M plan be linked to any other types of plans?

g) Review of costs and benefits of operation and maintenance

The data collected from work orders can be used to examine various activities to see whether benefits are accruing from the work. Some examples include:

- Maintenance activities that show the highest cost per year or the highest cost per event—Are these costs/efforts resulting in a positive effect on performance of the system, an asset or an asset class?
- Maintenance on highest risk assets—Have these types of maintenance activities prevented catastrophic events? Have they allowed assets to remain in service longer?
- Maintenance tasks the individual operators or maintenance personnel believe are worth investigating (i.e., an operator or maintenance person believes there may be a better way or time frame to perform a particular task on an asset)—Will changes in maintenance have a positive effect on the asset or class of assets?
- The most frequently performed maintenance tasks over the course of a year—Are these tasks ones that should be the most frequent? Does this make sense? Are the benefits worth the time and monetary costs of the activities?
- Maintenance on assets that have the most frequent failures—Should the system examine the potential for an "improve don't just fix it" approach for these assets? What kinds of changes could be made to decrease the number of failures?

The goal is to determine the most efficient and effective way to keep the assets in operation delivering the required level of service with the fewest failures and/or to extend the asset's life beyond what would have occurred.

h) Using data on asset interventions to predict asset condition, failures, and changes to future maintenance.

VIII. LONG-TERM FUNDING

This section includes information related to funding the full costs of operation of the facilities and how that will be accomplished.

a) Budget process
b) General sources of funding
c) User rates and fees
d) Definition of full costs of operations—Are these being fully funded?
e) Estimated replacement cycle
f) Costs of deferred maintenance
g) Equivalence of operational and capital dollars

h) Capital improvement planning process—Short- and long-term
i) Business case evaluation process
j) Prioritization process for capital projects
k) How capital projects are funded
l) Annual expenditures on capital projects (current and future)
m) Affordability concerns and how to address them

IX. STAFF/STAFFING CONSIDERATIONS

This section includes information about the staff and staffing aspects of the organization, including current staffing levels, future staffing needs, and related items.

a) Governance of the organization
b) Organizational chart
c) Staffing levels when fully staffed
    a. Operational
    b. Capital
d) Current and anticipated vacancies
    a. Percentage eligible for retirement
    b. Succession planning
e) Methods to fill vacancies
f) Three-part partnership of organization—Governing body, customers, staff
g) Benefits expected for staff, governing body, and customers

    All three of the partners need to see benefits from asset management. Some benefits are universal, such as protection of public health and the environment which benefits all three categories, while others are more specific to one category.

Appendix B

Teaching Asset Management in Engineering (Or Other Water Related Major): Creating a Class or Portion of a Class at the University Level

Asset management as a concept and a way of doing business is growing in practice for water utilities. There are regulatory requirements for asset management within several states and some state and federal funding agencies have requirements for asset management to obtain funding for capital projects. In other cases, states and federal entities have incentives for asset management to encourage the adoption of the practice. These regulations and incentives are likely to increase over time and enable more utilities to undertake and implement asset management. Because the engineering profession plays an important role in designing systems, meeting regulatory compliance, assessing asset condition, and helping systems problem solve, it is important that the engineers be partners with systems when it comes to asset management. In some cases, the engineers may help a system with portions of asset management development or implementation. Therefore, having knowledge of asset management can be important for new engineers and those returning to school for additional education or advanced degrees. Those individuals with knowledge and skills in asset management may be sought after by employers who are working to implement asset management with the clients they serve.

To receive all the benefits from asset management practice, it is necessary for all sectors (regulators, funders, utilities, elected leaders, engineers) to understand and engage in the practice. There are also significant advantages to the engineering profession when systems engage in asset management and implement the practice:

- Systems with better mapping and an inventory of assets, including asset attribute information, are easier to upgrade or revise in some way. In these cases, much less effort is required in the preliminary data gathering phase.
- A better understanding of criticality and useful life remaining leads to easier phasing of projects.
- Systems that fully embrace asset management understand the need to replace assets and will be seeking engineers to do the design work.
- While some projects may be smaller in scope if systems invest on a more incremental basis, there may ultimately be more work overall because systems would recognize the need for investments in assets.
- Systems engaged in asset management should be easier to work with as they will have a greater understanding of the need to actively engage with professionals as well as a greater understanding of the system.

It will be necessary for engineers to either learn about asset management on the job or in the classroom. It is not a matter of if engineers learn this information but rather when and how they learn it. Ideally, the learning should start at the university level before graduation.

There are several options when it comes to building asset management into the curriculum of an engineering program at the university level.

- A semester-long course in asset management
- A short series of classes (2 to 4) on asset management built into design courses on water or wastewater
- A single class focused on asset management
- A project for seniors that involves helping a small water system with asset management activities or creating an asset management plan for a small water system

Each of these options will be described in more detail below. Ideally, the curriculum could include a semester-long class followed by a project for seniors perhaps in the final semester. A single class will only introduce the topic but will not enable the students to fully understand asset management. It may be helpful for the students to have some prerequisites before taking asset management, such as engineering economics or some public works finance and accounting. There could also be some supplemental reading to aid in understanding of these concepts in lieu of a prerequisite.

A Semester-Long Course in Asset Management: Because there are so many aspects to asset management, ideally an entire course could be created. At this time, it is most likely that a course of this type would have to be an elective. Over time, it is possible that accreditation agencies may realize that this is information that at least civil engineers need to know and maybe it could be added in the future as a required course.

It is assumed that a college semester is between 14 and 18 weeks depending on the school. For the purposes of this book, it is assumed to be 16 weeks with classes meeting 3 days a week for approximately 50 minutes per class. Table B.1 contains an outline for a semester-long class.

A short series of classes (2 to 4) on asset management built into design courses on water or wastewater: If the option of teaching a small number of classes within a curriculum on water or wastewater treatment is used, the class topics can include what is shown on Tables B.2 and B.3.

A single class focused on asset management: If a single class it taught on asset management the topics to include are shown below.

- What is asset management?
- Why do engineers need to know about asset management?
- What do engineers need to know?
- Working *with* a utility on asset management not for a utility
- Triple bottom-line benefits to engineers and utilities related to asset management

A project for seniors that involves helping a small water system with asset management activities or creating an asset management plan for a small water system: Instead of adding asset management to the curriculum of a class or creating an entire class around asset management, another option is to have seniors work with water systems to implement portions of asset management and/or to create an asset management plan, depending on the amount of time available. It would probably be easier to work with a smaller system that had fewer assets and less complexity, but any size utility could work. There are tens of thousands of small systems across the country, so it would be easy to find a nearby water system in any state in the country and in urban and rural areas.

Some examples of activities that could be completed for small water systems include the following:

- A map of field assets
- An asset inventory or an inventory for a class of assets
- Level of service goal development

TABLE B.1 Semester-Long Course on Asset Management

Week/Class Period	Topic/*Description*	Homework
Week 1: Class 1	Introduction to Asset Management *This class is focused on Chapter 1: What asset management is. It should be a general introduction to the concept of asset management and the benefits it is intended to provide for a water, wastewater, and stormwater facility. It should discuss the fact that asset management is a thinking process not a technology process. It should also introduce the concept of why an engineer (or other professional if not an engineering class) should care about asset management. It can also be used to discuss the basic class expectations.*	Read Chapter 1. Be able to define asset management in your own words
Week 1: Class 2	The Five Core Concepts of Asset Management *A description of the five core concepts of asset management and how they fit together. Describe the important elements of each of the core concepts.*	Read Chapter 2 in preparation for Level of Service Class.
Week 1: Class 3	Level of Service: Mission Statement *What is a mission statement? What is it used for? How do mission statements help the organization align? What are good and bad elements of a mission statement?*	Research mission statements for water and wastewater systems on the web. Choose two mission statements from water utilities and two from wastewater utilities and compare them to the good and bad elements of mission statements. How well do they compare? Be ready to share with the class.
Week 2: Class 1	Level of Service: Mission Statements Leading to Goals *Discuss how mission statements lead to goals. Have students get into small groups and share the mission statements they found on the web. Each group should choose one of the water mission statements and one of the wastewater mission statements and determine what types of goals would be important to meet those two mission statements. This class should describe what level of service goals are and the fact that goals have service-level bands around them.*	Take the mission statements the students found on the web in the previous homework assignment and have them describe goals related to their water and wastewater mission statements to repeat the activity done in class (if their individual mission statements were not used in the class assignment). If their mission statements were used, ask the students to try to come up with additional goals beyond those discussed in workshop during the class.

(*continues*)

TABLE B.1 Semester-Long Course on Asset Management (*Continued*)

Week/Class Period	Topic/*Description*	Homework
Week 2: Class 2	Level of Service: External and Internal Goals *This class should begin digging deeper into what level of service goals are and how external and internal goals relate. The goals selected in the prior class are most likely going to be external goals. What are internal goals that would have to be created to meet the external goals selected? Have a class discussion on a mission statement and the external goals that lead from the mission statement. Then move to internal goals that would be required to meet the external goals.*	Take the two mission statements and goals created from the prior class and determine what internal levels of service goals would be necessary to meet the external service levels.
Week 2: Class 3	Level of Service: Measuring Goals *It is imperative to be able to measure goals in order to understand if they have been met. Go through several types of goals and discuss the data needed to measure the goals. How would the data be obtained? How would it be stored? How would it need to be analyzed? How often would goals need to be measured? How does that change with different types of goals?*	Using the external and internal goals developed in the prior classes, determine what data is needed to meet each goal, how the goals would be measured using the data, how the data could be stored and analyzed, and how often each goal should be measured and why that time frame was chosen.
Week 3: Class 1	Level of Service: Adjusting Activities to Meet Service Goal Targets *This class will discuss actions that can be taken when a level of service goal is not met, including adjusting the goal or target level and changing practice in some way to meet the target in the future. This class should include a variety of level of service goals and various actions that might be taken.*	Select two goals the students have been working on and have them imagine scenarios in which the goal was not met. What actions could be taken to allow the goal to be met. What changes to the goal or target level might also be considered? Is this a goal with service level bands that are large enough to allow a change in target levels? What are the implications of changing the target levels for the goal?
Week 3: Class 2	Current State of the Assets: What Is an Asset? *A discussion regarding what an asset is. What types of criteria define an asset and how assets are needed to meet the level of service goals established. It wouldn't be possible to meet goals if there were no assets. What types of assets make up water, wastewater, and stormwater facilities. Green assets versus gray assets (Chapter 7) can also be discussed.*	Read Chapter 3. Develop a flow chart that can be used to determine what an asset is.

(*continues*)

TABLE B.1 Semester-Long Course on Asset Management (*Continued*)

Week/Class Period	Topic/*Description*	Homework
Week 3: Class 3	Current State of the Assets: Condition Assessment *This class should focus on what the condition of assets means and how the condition of assets can be rated.*	Develop a rating system for a class of assets of the student's choosing from water, wastewater, or stormwater.
Week 4: Class 1	Current State of the Assets: Useful Life Remaining *This class should focus on what the useful life remaining is and why it is not the same as age of the asset. Age is a simplistic and generally underestimation of useful life. Describe how it can be estimated by utility personnel. Also discuss how an outside professional (such as an engineer) could aid in the useful life remaining estimation.*	Develop a method of estimating useful life remaining for the class of assets that was selected for the condition assessment. Describe how useful life remaining can be used in the practice of drinking water, wastewater, or stormwater design projects.
Week 4: Class 2	Current State of the Assets: Mapping Assets *Describe the mapping process as discussed in Chapter 3. Talk about ways of making maps and the fact that many different types of maps can be helpful. Discuss that there are free ways of creating maps.*	Investigate three different software programs that can be used to create a map. What would each cost, what would each entail in terms of annual fees (if any), what type of training would be required to use them, what are the capabilities of the mapping system, and what size/type of system would best be able to use it.
Week 4: Class 3	Current State of the Assets: Collecting Data Using a Data Collection App *Describe the need to determine what attributes should be collected on assets. Which are required, which are optional. Pros and cons of collecting more or less attribute information. Discuss ways to collect data (attributes) on assets during the field asset inventory process.*	Determine what attributes should be collected for the class of assets that have been used for the prior class homework assignments (starting in Week 3 Class 3). Develop a Data Collection App using whatever approach the student chooses.
Week 5: Class 1	Current State of the Assets: Collecting Data in the Field *This class should be outside using visible assets near the university, such as meters, hydrants, manholes, valves, or any other assets that are nearby. The students should work in groups, using their collection apps to collect data on the assets (including location data and as many attributes as can be collected in the field).*	Describe the challenges and successes of the data collection process. What worked, what didn't? Would you need to modify the app in any way after trying it? If so, describe the changes. What data was able to be gathered in the field, what data would need to be gathered by operating the asset or by looking at other data sources that would be more likely in the office somewhere? Go around campus or another neighborhood and map at least 20 assets.

(*continues*)

TABLE B.1 Semester-Long Course on Asset Management (*Continued*)

Week/Class Period	Topic/*Description*	Homework
Week 5: Class 2	Student Activity Presentations: Each student should present their map *Every student should show their map to the class and discuss the answers to the questions in the homework regarding asset data collection. If there are too many students for each to speak during the class, students could be selected at random to present.*	None
Week 5: Class 3	Current State of the Assets: CMMS and Asset Management Programs *Include a discussion of CMMS versus Asset Management Programs, along with the differences. It should also include a discussion of why a system should want a system and the process of selecting a software.*	Investigate one CMMS or Asset Management Program. Describe whether it is an AM or CMMS system, what the program is intended to provide, who it could be used by, what it would cost (initial and annual fee), how much training is required, whether it works out of the box or has to be customized, and what functions it provides.
Week 6: Class 1	Student Presentations on CMMS. Choose several students to present on the CMMS software they researched. Discuss the pros and cons of the different software programs as a group activity after the presentations.	None
Week 6: Class 2	Criticality: Probability of Failure *A discussion of probability of failure, including the 4 failure modes and how probability of failure is rated.*	Read Chapter 4.
Week 6: Class 3	Criticality: Consequence of Failure *A discussion of consequence of failure, including the triple bottom-line aspects of consequence of failure and how consequence of failure is rated. This should include a discussion of how some of the factors may be monetized and some not.*	Develop a means of assessing the probability of failure and consequence of failure for the type of assets that were used in the previous homework assignment (Week 3, Class 3). Choose 10 assets of that type in the local area and conduct a POF/COF analysis. Use whatever data is available and supplement with any assumptions the student wishes to make. The assumptions should be written out and included with the analysis. For students choosing non-visible assets or those in buildings, work with the students to structure an approach to address this type of asset.

(*continues*)

TABLE B.1 Semester-Long Course on Asset Management (*Continued*)

Week/Class Period	Topic/*Description*	Homework
Week 7: Class 1	Criticality: Visualizing Criticality *A discussion of why visualizing criticality is important as well as examples of ways criticality can be visualized.*	Design a way to visualize the set of criticality data developed in the previous homework. It can be one of the ways discussed in class or any other way of the student's choosing.
Week 7: Class 2	Life Cycle Costing: What is an asset's life cycle? *A discussion of the life cycle of an asset including a discussion of why starting the asset's life earlier than initiation of operations is so important. A discussion of the basic interventions that can be done along the life cycle and how these interventions are meant to prolong the life of the asset and ensure its proper operation.*	Read Chapter 5.
Week 7: Class 3	Life Cycle Costing: How engineers should work with operations personnel during design *A discussion related to the ways in which an engineer and operations personnel can work together to have the best design and operations. Reasons why this relationship is important and how it can help both sides, including ensuring both parties are speaking the same language and understanding each other should also be included.*	List 5 ways you, as an engineer, could work with water utility personnel.
Week 8: Class 1	Life Cycle Costing: The relationship between construction and asset performance *Construction techniques and modifying these techniques is an overlooked part of asset management. It is necessary to understand how construction relates to an asset's life and if that construction can be performed in a better way, the asset's life may be less problematic and increased.*	Consider the construction techniques that may be used for the set of assets you've been working on (Week 3, Class 3). Which techniques may relate to the asset's life? In what ways? How might they be revised or what standards might be set to increase the asset's life? For example, if pipe is laid with bedding sand but many breaks related to punctures are occurring, bedding sand may need to be increased.
Week 8: Class 2	Midterm *This test should include: Introduction to Asset Management, Level of Service, Current State of the Assets, and Criticality. Life Cycle Costing should be in the final but not the midterm.*	None

(*continues*)

TABLE B.1 Semester-Long Course on Asset Management (*Continued*)

Week/Class Period	Topic/*Description*	Homework
Week 8: Class 3	Life Cycle Costing: Operations and Maintenance *A discussion of operation and how operation can affect asset life, the need for repairs, and trying to maximize uptime. A discussion of the reasons for maintenance, the different types of maintenance and how they relate to asset criticality.*	Describe the operations related to the asset class the student has been working on (Week 3, Class 3). How does the operation relate to asset life, need for repairs, or uptime? Is there any modification to the operation that would help with asset life? Describe at least 1 type of maintenance for each category for the asset class: routine, preventive, predictive, corrective. If there are no activities in that category, describe why not.
Week 9: Class 1	Life Cycle Costing: The Concept of Don't Just Fix It, Improve It *One option when an asset fails repeatedly is to just continue to fix it over and over again. A better strategy is to actually change the situation in some way so the repair is no longer needed. For example, a pump that continues to vibrate may need many replacement parts. Rather than replacing the parts, the pump alignment (which is causing the vibration) could be fixed, eliminating the vibration altogether.*	Imagine a type of event that could repeatedly occur on your class of assets. What type of improvement could be made to prevent that from occurring in the future? Be as creative as possible.
Week 9: Class 2	Life Cycle Costing: Repair, Rehabilitate, Replace *When assets fail, there are three basic options: repair, rehabilitate, and replace. This class will discuss which option should be chosen under which conditions and how this decision can relate to criticality. It should also include a discussion of run to failure as a management strategy and which assets might be best for this type of strategy.*	Consider the asset class the student has been working on. What repair, rehabilitation, and replacement options exist for those assets? Would this set of assets be a good candidate for a run to failure strategy, why or why not?

(*continues*)

TABLE B.1 Semester-Long Course on Asset Management (*Continued*)

Week/Class Period	Topic/*Description*	Homework
Week 9: Class 3	Life Cycle Costing: Capital Improvement Plan *A description of what a capital improvement plan (CIP) is and how one can be created out of the asset inventory. A description of the difference between a more detailed short-term CIP and a less detailed long-term CIP. Why a long-term CIP is important for planning purposes.*	Consider the estimated useful life for the asset class the students have been working on. This was something the students would have completed in the Current State of the Assets homework assignments. Estimate a replacement cost for this class of assets or the individual assets within the asset class. Develop a short- and long-term CIP for these assets. Short term is 10 years or less, while long term would be up to 50 to 100 years. If the assets don't last more than 50 years, how many times would they need to be replaced within the 50 to 100 year time frames.
Week 10: Class 1	Life Cycle Costing: The Business Case Evaluation *This class should discuss when a business case evaluation should be used and what it includes.*	Develop a business case evaluation to replace an asset or collection of assets within your asset class.
Week 10: Class 2	Long-Term Funding: User Rates *This class should discuss how user rates are used to pay for most expenses within the water utility. The goals that are imbedded within rate structures and how these goals can be contradictory to one another should also be discussed.*	Read Chapter 6.
Week 10: Class 3	Long-Term Funding: Debt Financing *This class should discuss the use of outside debt financing to replace assets as well as the types of agencies that can fund projects and how the various funding sources work. This class should also describe some of the complexities of funding and why some utilities may choose to use other funding sources such as bonds or private banks.*	Choose one funding source for which a utility wishing to replace the class of assets might apply. Investigate the funding cycle, the interest rate, the payback period, and the additional requirements (e.g., engineering reports, environmental documentation, "Build American Buy American," etc.). State whether you think this funding source would or would not be a good option for asset replacement.

(*continues*)

TABLE B.1 Semester-Long Course on Asset Management (*Continued*)

Week/Class Period	Topic/*Description*	Homework
Week 11: Class 1	Long-Term Funding: Incremental Funding *Describe the concept of incremental funding and why this would be important for utilities to consider. Describe the ways in which incremental funding could be used on different asset classes. Which ones are easier or more difficult? Have students share how they think incremental funding would work with their class of assets.*	None
Week 11: Class 2	Long-Term Funding: CapEx vs. OpEx and How It Affects Asset Management *Describe how Capex and OpEx are funded differently and therefore can have an effect on how the utility practices asset management. The utility may choose to weigh CapEx more heavily than OpEx because they fund OpEx but outside funding can be used for CapEx. There may be grants for CapEx but not for OpEx.*	List CapEx and OpEx expenditures that might occur with the student's asset class, such as operation and maintenance activities, and rehabilitation or replacement activities.
Week 11: Class 3	Green Asset Management *Green assets can be included in an asset management plan. Describe what assets can be included for different types of utilities (water, wastewater, and stormwater) and why it can be important to include them. Describe the parts of asset management that are more difficult for green assets than gray.*	Read Chapter 7.
Week 12: Class 1	Asset Management Plan: What Is It and What Does It Include? *This class should include a discussion of asset management plans, what they include and how they should be written and the fact that they should be able to last for a period of time (up to 5 to 10 years) without revision.*	Read Chapter 8.
Week 12: Class 2	Sustaining Asset Management *It is necessary to sustain an asset management program over time. This class should discuss ways that such a program could be sustained. The class can also discuss why it is difficult to sustain a program over time.*	Come up with two or three ways a utility can sustain its asset management program. How difficult would these ideas be to implement?

(*continues*)

TABLE B.1 Semester-Long Course on Asset Management (*Continued*)

Week/Class Period	Topic/*Description*	Homework
Week 12: Class 3	Measuring Progress in Asset Management *Discuss the use of the asset management IQ process to measure progress in asset management. Discuss why measuring practice is an important activity to undertake.*	Go onto the SW EFC website and look at the AM IQ tool and what it entails. Fill out answers to the questions (you can choose the answers you want to each question but must vary the responses from 0 to 5) and print out the results to turn in. Describe what those results might mean for the utility. What would it take for the utility to increase their score 10 points above where they are now? What specific activities could they take, given the answers you input into the software?
Week 13: Class 1	Measuring Triple Bottom-Line Benefits of Asset Management *Describe the triple bottom-line benefits of asset management and how these benefits can be assessed (some are quantitative, and some are qualitative). Discuss why it is important to measure benefits.*	Describe expected benefits that may occur if asset management was practiced on the class of assets the students have been working on. Which are more easily quantified? For those that are more qualitative in nature, how can quantitative measures be included in the analysis?
Week 13: Class 2	Discussion of Class Presentations *This class should describe how students will create a class presentation out of the homework they have done during the semester. It should also discuss how to present the information in a professional manner and show examples of good presentations. It can include discussion time for students to ask questions and share in a peer-to-peer manner in small groups.*	Work on presentations.
Week 13: Class 3	The Human Element of Asset Management *Describe the human elements of asset management and why that is an important aspect of asset management. Discuss how some of these elements can be assessed (metrics to assess them).*	Read Chapter 9.
Week 14: Class 1	Communicating Asset Management *It is necessary to communicate with elected leaders and customers on asset management activities. Discuss ways that this can be accomplished. What are the pros and cons of different ways of doing this?*	Develop one method each of communicating asset management activities and/or successes with the elected leaders and the general public. How well do you think these methods would work?

(*continues*)

TABLE B.1 Semester-Long Course on Asset Management (*Continued*)

Week/Class Period	Topic/*Description*	Homework
Week 14: Class 2	Understanding Asset Management Terminology Class Activity *This activity would be best done using some type of polling software. The class should include putting a word on screen and then having several definitions below it. The class should use polling software to pick the best answer. The intent is to see how well asset management terminology is embedded within the class. If most people are able to answer the questions correctly, the terminology is fairly well imbedded. If not, it is not understood as well as it should be.*	Choose five words related to asset management and define them in your own words.
Week 14: Class 3	Benefits of Asset Management to Engineers *Small group discussions during class to discuss the ways in which engineers can benefit from having a community develop and implement a robust asset management program. For this exercise, we are interested in the benefits* ***to the engineer,*** *not the system.*	Choose one benefit for engineers that you think would be the most important to you personally in your career. Describe why you believe this would be the most important one.
Week 15: Class 1	What an Elected Leader Needs to Know About Asset Management *This class will discuss what an elected leader needs to know about asset management. Their need to know is less than operators and managers given their role in the system, but it is still important that they understand the basics.*	Read Chapter 10.
Week 15: Class 2	Working With a Utility Not for a Utility on Asset Management *An asset management plan will not be effectively implemented if the plan is completed for the utility instead of with the utility. The utility must be part of the process. This class can include a discussion of why that is so valuable in not just asset management but in other ways the engineer might work with a community. This class should include discussion.*	Prepare for Presentation.

(*continues*)

TABLE B.1 Semester-Long Course on Asset Management (*Continued*)

Week/Class Period	Topic/*Description*	Homework
Week 15: Class 3	The Regulatory Environment *Describe the current regulatory environment in the state in which the class is being taught. What is required by regulation, by funding, or some other way for asset management? What incentives are there? How do these work? What funding is available in the state to support asset management at a utility?*	Prepare for Presentation.
Week 16: Class 1	Class Presentations *Each student will present on the work they did for asset management for their class of assets. The presentations should be 10 to 15 minutes in length. They should use professional practices to make the presentation. They should be graded on both content and professionalism. Note: This course was designed with the students doing the projects individually. It is also possible that the students could work in small groups to do the activity. In this case, these would be team presentations. It is also possible to have the students sign up for assets at the beginning of class to ensure that only one student or group of students selects an asset class. It is also possible to choose only certain asset classes and have the students select one of those assets.*	Prepare for presentation.
Week 16: Class 2	Class Presentations *Continue Class Presentations*	Prepare for presentation.
Week 16: Class 3	Class Presentations *Continue Class Presentations*	Prepare for presentation.
	Class Final *The final should include all material but focus more heavily on the second half of the course.*	None

TABLE B.2 Four Classes on Asset Management

Class Period	Topic	Homework
Class 1	Introduction to Asset Management and Level of Service *This class is focused on Chapters 1 and 2, including what asset management is and what it is meant to provide for a water, wastewater, and stormwater facility. It should discuss how asset management is a thinking process not a technology process. It should also describe what the mission statement is and how it relates to level of service goals, including how to set and measure LOS goals.*	Read Chapters 1 and 2. Choose an asset class and write two goals that would be relevant to that asset class.
Class 2	Current State of the Assets and Criticality *This class is focused on Chapters 3 and 4, including what an asset is, what attributes can be collected on the asset, and how a map can be made. It should also discuss probability of failure, consequence of failure and how to visualize criticality.*	Read Chapters 3 and 4. Determine what attributes you would want to collect for the class of assets from Homework 1.
Class 3	Life Cycle Costing and Long-Term Funding *This class will focus on Chapters 5 and 6 including the life cycle of an asset and the types of interventions that can be completed for each stage of the asset's life. It should discuss how operation, maintenance, repair and replacement relate to criticality. The class should discuss how funding can be obtained (rates and outside sources for asset replacement), the concept of replacement cycle, and how a capital improvement plan can be created.*	Read Chapters 5 and 6. List at least one intervention that could be done on the class of assets selected in Homework 1 for each phase of the asset's life cycle.
Class 4	Implementing Asset Management, including Asset Management Plans *Discuss the need to implement asset management, including writing an asset management plan and what it takes to sustain an asset management program over time.*	Read Chapter 8.

TABLE B.3 Two Classes on Asset Management

Class Period	Topic	Homework
Class 1	Introduction to Asset Management and the Five Core Components *This class is focused on what asset management is and what it is meant to provide for a water, wastewater, and stormwater facility. It should discuss how asset management is a thinking process not a technology process. It should include the 5 core components of asset management and the basics of what each entails.*	Read Chapter 1.
Class 2	Implementing Asset Management, including Asset Management Plans and Measuring the Benefits of Asset Management *Discuss the need to implement asset management, including writing an asset management plan and what it takes to sustain an asset management program over time. This class can also include a discussion of the human aspects of asset management.*	Read Chapter 9.

- Design of a measurement tool to measure level of service goals
- Design of a communication tool to show how well the goals have been met
- Criticality analysis
- Criticality visualization
- Development of standard method to assess probability of failure and consequence of failure
- Creation of an operation and maintenance plan
- Development of a business case evaluation process
- Calculation of a replacement cycle
- Review of needed funding

If the project is intended to last an entire year, it can also include the development of an asset management plan that summarizes all of the activities. It would be best if students worked together in teams. This activity could also be a supplement to a semester-long class in which the first semester is the class and the second semester is a project of this type.

While this chapter focused on building asset management into the curriculum of an engineering program, there are other majors that could benefit from learning about asset management, such as environmental science, water resources management, sustainability, or any other majors that touch on drinking water, wastewater, and stormwater utilities. These majors can use any of the techniques discussed above and modify as necessary to fit specific needs.

Index

www.ingramcontent.com/pod-product-compliance
Lightning Source LLC
LaVergne TN
LVHW061732140726
843108LV00003B/6
* 9 7 8 1 5 7 2 7 8 4 3 4 5 *